Intensely Human

Intensely Human

The Health of the Black Soldier
in the American Civil War

MARGARET HUMPHREYS

The Johns Hopkins University Press
Baltimore

All rights reserved. Published 2008
Printed in the United States of America on acid-free paper
2 4 6 8 9 7 5 3 1

The Johns Hopkins University Press
2715 North Charles Street
Baltimore, Maryland 21218-4363
www.press.jhu.edu

Library of Congress Cataloging-in-Publication Data
Humphreys, Margaret, 1955–
Intensely human : the health of the Black soldier
in the American Civil War / Margaret Humphreys.
p. cm.
Includes bibliographical references and index.
ISBN-13: 978-0-8018-8696-6 (hardcover : alk. paper)
ISBN-10: 0-8018-8696-1 (hardcover : alk. paper)
1. United States—History—Civil War, 1861–1865—Participation,
African American. 2. United States—History—Civil War,
1861–1865—Health aspects. 3. United States—History—
Civil War, 1861–1865—Social aspects. 4. Russell, Ira, 1815–1888.
5. African American soldiers—Health and hygiene—History—19th
century. 6. African American soldiers—Mortality—History—19th
century. 7. Masculinity—United States—History—19th century.
8. Body, Human—Social aspects—United States—History—
19th century. 9. Racism—United States—History—19th century.
10. United States—Race relations—History—19th century. I. Title.
E540.N3H86 2007
973.7'0896073—dc22
2007013962

A catalog record for this book is available from the British Library.

Special discounts are available for bulk purchases of this book.
For more information, please contact Special Sales at 410-516-6936
or specialsales@press.jhu.edu.

For Will

CONTENTS

"Anyone can do any amount of work, provided it isn't the work he is *supposed* to be doing at that moment."[1] Thus Robert Benchley advised his readers in an essay on procrastination and accomplishment. He had five tasks to do, and he put writing a newspaper article at the top of the list. He spent the rest of the day not working on the article, but he finished the other four tasks with splendid ease. The pages that follow this preface are one more illustration of how much you can get done if you work hard enough at not doing the assignment boldly printed (and underlined in red) at the top of your list.

In my case, the official task was writing a history of medicine in the U.S. Civil War, and, more particularly, the war's effect on American medicine. I had grants with which to do that; I had a sabbatical during which to do that; I had a lovely office in the National Humanities Center in which to do that. But I haven't done it—at least not yet. Instead, that envisioned book's chapter on black soldiers nibbled that seductive cake labeled "eat me" and morphed into a monograph all its own.

A major distraction from my stated purpose was Ira Russell, a Civil War physician whose writings abound in the papers of the United States Sanitary Commission, a voluntary organization that functioned something like the American Red Cross would in later wars. Russell was a compassionate physician who became engrossed with the cause of the black Civil War soldier and his health. He had first cut open a black body in 1843 or 1844 as a medical student in New York City, where many of the bodies intended for dissection were shipped in barrels from the South. After returning home to New England and settling in Natick, Massachusetts, he probably saw few African Americans in his practice. But he was no doubt aware of the rising abolitionist sentiment roiling the state, and he became friends with Henry Wilson, later a firebrand senator from Massachusetts who defended the rights of black people during and after the war. When men from Natick formed a regiment in response to Lincoln's call for troops, Russell joined as well. The medical department ultimately promoted him to running general hospitals, first in Arkansas and later in St. Louis. There Russell watched as trainloads of black men arrived, ragged, hungry, and half frozen, to be mustered into newly forming black regiments. Russell watched, too, as the men grew feverish, needed hospital beds, and were denied them by a "Copperhead quartermaster." He wrote to Wilson, who shamed the army on the Senate floor for its inhumane treatment of its newest soldiers. Russell had acquired a new vocation: the care, defense, and study of the black soldier's body.[2]

Letters and reports by Russell and others reveal a deep fascination with the workings of the black body among northern physicians newly exposed to black men in hospital and camp.[3] The Civil War brought the fate of African Americans to the center of American consciousness. There were some 4 million slaves in the South; by 1863 it was clear that these people might soon be free and perhaps heading for the northern states. Would they be law abiding? Would they work without coercion? Were they healthy enough and smart enough to survive on their own? These questions were intense and immediate for many northerners, especially ones who had seen little of these dark-skinned people in the years before the war.

Russell and like-minded physicians saw in the vast experiment of recruiting black men into the Union army an opportunity for scientific study and medical progress. The Sanitary Commission surveyed doctors who had treated large numbers of black soldiers, inquiring into areas of difference. Black soldiers were measured, weighed, and tested when they left the army and were compared to white men. With great interest, northern physicians, including Russell, dissected the black body for clues as to its distinctiveness. And they had many bodies to explore, for poor treatment led to high rates of disease and death among black troops. Russell loudly protested the mistreatment—and was overtly proud of his autopsy work.

This book follows Russell and his colleagues as they explore, exploit, and explain the black bodies made available to their view by the enrollment of more than 180,000 black men in the Union army. Other historians, notably Joseph Glatthaar and Paul Steiner, have described the high mortality rates black soldiers suffered due to disease during the war, and chronicled their shabby treatment at the hands of the Union army.[4] This narrative builds on the work of these and other scholars but uses new archival materials to deepen our understanding of northern physicians' view of the black body, of its strengths and weaknesses. It draws on nineteenth-century tropes of manhood, racial mixing, and bodily strength to understand how men likened to brute animals such as oxen could simultaneously be considered lacking in endurance. My account, especially in chapters 2 and 8, emphasizes how these arguments about the black body were used within the broader debates about the place of the black man in the strange new world that would emerge from emancipation and the war.

The reasons for high black mortality rates, described briefly by earlier historians, are made clearer here by close studies of particularly morbid environments in St. Louis, South Carolina, Louisiana, and Texas. Chapters 4, 5, 6, and 7 indict the Union army and its medical department for poor management and deliberate malfeasance in the care of black soldiers. These chapters also emphasize how little black troops could advocate for themselves. Many white men were literate and could write home to family members or to the local newspaper to complain if conditions were bad. White volunteer regiments were often officered by men from the same region as the foot soldiers. Because these officers were bound to their men by geographical ties, they had reason to advocate for them, whether by asking the quartermaster to deliver better quality goods or calling on the Sanitary Commission for supplements of food, clothing, or blankets. Privates in the black regiments had very limited skills for self-advocacy, and their white officers often had few bonds of affection or loyalty. The black man suffered as a result. The stories told in chapters 4 through 7 demonstrate the importance of these issues in determining the health of a regiment, while also heralding the influence of a few literate black men and the compassion of a few white advocates for the black troops.

Chapter 3 and sections in other chapters explicate the epidemiological patterns beyond the control of nineteenth-century physicians which led to high black mortality from disease. Central to understanding these patterns is the epidemiology of the major killing diseases, an epidemiology at times complicated by the history of racial groups' prior disease exposure. These chapters could be written entirely from the perspective of nineteenth-century physicians, a perfectly justifiable approach that sidesteps the question of whether these physicians were right about any of the disparities they noted between black and white patients. Yet I prefer to take seriously the possibility that health disparities among black and white troops might in part have depended on biological differences, and I explore this question with an open mind. Modern research in this field is complicated by claims that race and ethnicity have no biological basis, that they are categories entirely socially constructed. These claims have in turn spawned an active debate over the use of these categories in medical research.[5] Chapter 3 dissects this contemporary debate and then considers the mid-nineteenth-century discourse on certain key diseases.

The book concludes with a discussion of how the black soldier's health experience was twisted and interpreted after the war to reinforce existing stereotypes and direct blame away from the white hierarchy that created the conditions in which disease thrived. Russell, who lived into the 1880s, is strangely silent after war; others used his information in the decades after the war to tell a tale that must have made him cringe. Perhaps once back in Natick, he wanted only to put the war experience behind him. The epilogue briefly considers the black soldier's experience following the war, drawing largely on Civil War pension records.

Many histories of Civil War medicine focus on the battlefield and the gory events that immediately followed the volleys of shot and shell. Regarding white soldiers, much has been written about ambulance organization, field hospitals, and care at more general hospitals. Much of that information is missing for black soldiers. First, black regiments participated in many fewer battles than whites, as the military hierarchy felt they were better deployed on garrison duty than the front lines. Second, much of what we know about white soldiers' experience comes from diaries, letters, and official reports, which are scarce for black troops. The men themselves were largely illiterate, and their officers were less likely to give written attention to the medical issues of black men. The story that follows has been pieced together at times from thin scraps—a letter in the National Archives here, a newspaper column there—often held together by the strong mesh of the Sanitary Commission documents. While it may lack the thickness of some accounts of the white soldier, it is a story that deserves telling, patches and all.

Black soldiers in the American Civil War were far more likely than their white comrades to die of disease. Many of them entered the war disadvantaged by a lifetime of malnutrition and the immunological naiveté engendered by their rural pasts. Others suffered deprivation in the days and weeks just preceding induction, making them even more susceptible to disease when crowded into camps. Much of this mortality, however, can be traced to specific choices made by officers, bureaucrats, and other authority figures, who proved poor stewards of the men in their care. Their decisions, great and small, careless and deliberate, doomed these soldiers to early graves.

A word about terminology is appropriate here. During the 1860s polite labels for persons of African heritage included *negro, black, colored,* or *African.* Initial names for black regiments were the Corps d'Afrique or the [number] [state name] Regiment A.D., with A.D. standing for "African Descent." After May 1864 all black regiments were labeled U.S. Colored Troops, with a regimental number. Most were infantry, and the abbreviation U.S.C.I. was also employed. Although the word *colored* has fallen out of fashion in the United States today (and has other meanings in countries such as South Africa where it refers to a different population), I have kept it here as accurate to the time period. The word *nigger* was never polite, even in the 1860s, and represented a certain attitude of derogation on the part of the speaker. Accordingly, when it appears in quotations I have retained it, as the word reveals much about the perspective of the thinker. In my own descriptions, I have used the adjectives *black* and *African American* interchangeably. I realize these are contested terms but know of no better way to refer to the men who are the subject of this narrative.[6]

A word about numbers is also in order. The Civil War was a massive bureaucratic enterprise, with the Union government creating vast systems to track the soldiers in its employ. What other company, then or since, has managed nearly 2 million men who had to be fed, clothed, paid, and cared for when sick? The Union government created for each man a card that recorded his location during the war as well as his diseases when disabled and hospitalization site when relevant. An army of clerks kept each card up to date; the U.S. Colored Troops section alone employed more than four hundred clerks. Thus, at the end of the war, various government offices could transform this mountain of data into detailed statistical tables that included information about battle casualties, disease rates, and disability. But the system was not perfect. Large numbers of men went "missing." Some were captured and later surfaced in prisoner-of-war camps. Some were so mutilated by battle that their remains could not be identified. Some fell in wooded areas like the Wilderness, their bodies forgotten until their skulls shocked campers years later. Some simply deserted. As a result, various compilations at the end of the war do not always agree. For example, the surgeon general's numbers do not match those coming out of the adjutant general's office, the army's equivalent of a personnel office.[7]

Other historians have taken on the question of which numbers are right. In the case of the black soldiers, estimates of their total numbers range from 179,000 to more than 186,000. There are 209,149 names on the "wall of honor" surrounding the African American Civil War memorial in Washington, D.C.; it includes the names of sailors as well as soldiers.[8] I have chosen to use the figure of "about 180,000." There are further problems in trying to make sense of the disease and death rates reported by the surgeon general in the Medical and Surgical History of the War.[9] The first two statistical volumes appeared in 1872 and contain data frequently cited throughout my account. The tables provide the number for "mean troop strength" for each fiscal year ending on June 30. This allows a population against which to create simple descriptive statistics of disease-specific mortality rates. There are obviously multiple sources of error in these data. The contemporary diagnosis (remittent fever, say) may or may not precisely map onto a modern diagnosis (in this instance, of malaria). The historian can either dismiss the numbers as worthless or say that they are the best we have and use them with caveats in mind. I have chosen the latter approach, at least for considering cause of death. Reporting of death was more accurate than reporting of disease. The soldier's presence had to be accounted for one way or the other, whereas illness was less demanding in its bureaucratic signification. When soldiers died, their pay stopped, and their regiments' supply of food, armaments, tents, and other material was likewise affected. These factors were not as altered when the soldiers were merely ill. In studying the 65th U.S. Colored Infantry historian Paul Steiner worked backwards from known deaths from disease to the expected number of cases that should have generated these deaths. Even given very high case fatality rates, there were large discrepancies in the estimation of cases.[10] So I have largely ignored reported cases and relied, with skepticism, on the disease death rates.

One might suspect that whatever the multiple sources of error, given such large numbers the differences in white and black statistics would even out. One might think, for example, that just as many cases of disease would be missed for black soldiers as for white men. But there is reason to think that this was not the case. The men who sent in the sickness reports and filled out the cause of death on forms sent back to Washington were physicians, and the black regiments were infamously short of medical personnel. There is also reason to believe that the line officers of the black regiments were the least experienced in the war and probably, at times, the least careful about the state of their troops. Such physicians would have been less likely to fill the forms out, or fill them out properly, and such officers less likely to enforce regulations. The officers and physicians may have been less attentive to the men as individuals, as well. There is at least one report that officers would fill gaps in regiments in Louisiana with available men from nearby plantations. To save paperwork, they gave the new recruits the names of the recently dead. This practice is perhaps slightly less pernicious than it seems, for many of the men acquired last names for the first time upon entering the army. Some even changed their names after enlistment, deciding, for example, to abandon their owner's last name for the name of their father or some famous figure. Names were fluid in black regiments in a way not seen among white troops. All of this complicated record keeping.[11]

Some comments about "theory" may be useful to the reader. It will be obvious from this narrative and my previous books that I value information from modern medical science and employ it where useful to explain the past. I confess that I privilege the knowledge and epistemology of twenty-first-century medicine and science over other medical belief systems. Some use the phrase medical realist or historical realist to describe this approach, which can be contrasted to cultural relativism (all medical belief systems are equally valid, just as all systems of morality are equally valid) and to social constructionism (diseases have no independent reality but rather are constructed out of bodily symptoms by individuals living in a time and place to serve certain aims of that social milieu). Charles Rosenberg has promoted a modification of these views that posits a biological frame grounded in nature upon which societal meanings are hung; my position approximates his view with perhaps stronger emphasis on the methods and outcomes of scientific explanation. My point of view is also strongly influenced by ecology and evolutionary biology. These various perspectives are evident throughout this book.[12]

Finally, the research that underlies this book is obvious in the footnotes, but a few words about the reliability of the sources are perhaps relevant here. Most of the primary documents used were written by white, mid-nineteenth-century American males. Some were overtly hostile to the black man; others were overtly sympathetic. All shared assumptions about black-white differences that strike us as racist today. This perspective makes them suspect as reporters of the condition of black soldiers and their experiences during the war. The historian often has to work with such tainted sources, making the best judgment possible of each reporter's biases and reliability. Like other historians, I tend to judge manuscript documents to be more valuable than published sources, as such texts have not passed through the filter of an editor with ideas of propriety or importance. Where possible, I have included letters or other sources written by black soldiers, but there are precious few of them. In most southern states it was a crime to teach slaves to read and write, and the vast majority of black troops had been born into slavery. Most of these letters were published in one of few northern black periodicals; sometimes the author's name is a pseudonym. In a letter written from Texas in the summer of 1865, a fictitious author made claims that were described as lies by another black soldier in a later letter to the same publication. Only after finding significant corroboration of the pseudonymous author's description of affairs did I feel fully confident in believing him. Even sources as apparently straightforward as letters from black soldiers to the press can be misleading or of uncertain value.[13]

Of even more uncertain value are the many wartime diaries published in the decades following 1865, when wartime memoirs sold so well. Still, several diaries or autobiographies contribute largely to this narrative, despite concerns about selective memory or self-editing in the peace of postwar reflection. One such memoir gives this book its title. Thomas Higginson was a Massachusetts abolitionist chosen to be colonel of a black regiment formed of slaves from plantations on the Sea Islands of South Carolina in 1863. Higginson's commanding officer, Gen. Rufus B. Saxton, received the questionnaire circulated by the U.S. Sanitary Commission concerning the black soldier's peculiar susceptibility to disease and response to treatment. He discussed it with Higginson by a campfire one night.[14] Higginson recorded that "General Saxton, examining with some impatience a long list of questions from some philanthropic Commission at the North, respecting the traits and habits of the freedmen, bade some staff-officer answer them all in two words,—'Intensely human.' We all admitted that was a striking and comprehensive description."

My freedom to explore these many fascinating themes (instead of writing the book that I was supposed to be composing) owes much to my time at the National Humanities Center in 2004–2005. Funded by a Frederick Burkhardt Fellowship from the American Council of Learned Societies, I had the freedom to change horses in midstream, completely reorient my project, and take off in this unexpected new direction. My colleagues at the Humanities Center fed the project in many ways. Fellows at the center form impromptu seminars around subjects of mutual interest, and my group chose to explore the study of sexuality. We shared papers and critiqued each other's work. Chapter 2 began there, and I am especially grateful to my seminar colleagues—Lynda Coon, Pete Sigal, Bruce Kapferer, Kevin Ohi, Bruce Redford, Georgia Warnke, Julia Clancy-Smith, Israel Gershoni, and Tony Day—for the ways that seminar changed my perspective. Others at the center were influential as well—through interactions that ranged from single conversations to frequent discussions—especially Edward Curtis IV, Gregg Mitman, Geoffrey Harpham, Kent Mullikin, Michael Gillespie, Cara Robertson, Deborah Harkness, Tom Cogswell, Mary Favret, Lawrence Jackson, Lisa Lindsay, Joel Marcus, Andrew Miller, and Ding Warner. Thanks.

The staff of the National Humanities Center supported this project in many discrete ways. The librarians, Eliza Robertson, Betsy Dain, and Jean Houston, filled pages of interlibrary loan requests and used great ingenuity in finding some truly obscure sources. The computer guys, Phillip Barron and Joel Elliott, kept all the machinery humming. Karen Carroll worked over available text as a copyeditor, and the prose is better for it. Lois Whittington and Pat Schreiber kept us organized, fed, and generally pampered. It is hard to imagine a more fruitful setting for scholarship than this beautiful place in the Carolina woods, and I am grateful to all who sustained my work there.

This project has been supported by the National Library of Medicine publication grant program (G13-LM008350-02). I first learned of this opportunity when invited by Todd Savitt to serve on the review panel for the proposals, and I am grateful to the program's various officers over the years, including Susan Sparks, Valerie Florance, Hua-chuan Sim, and Zoe Huang, for their help. Susan Lasley in the Office of Research Support at Duke University has made sure that the grant renewal forms were done in a proper and timely manner, keeping the three-year grant flowing. Cynthia Hoglen and Vivian Jackson, administrators in the History Department at Duke, also bore with me as we together figured out how to complete the forms and calculate the necessary numbers.

Work on this book was indirectly influenced by participation in an in-house fellowship program at Duke under the auspices of the Social Science Research Institute. This seminar brought together scholars from law, sociology, psychology, economics, history and epidemiology to discuss topics related to contagion and social networks. The various quantitative sections of the book grew out of my experience with this colloquium. In turn, we generated a grant proposal to work with the dataset on Civil War veterans created by Robert Fogel's group at the University of Chicago. I am indebted to these colleagues—Frank Sloan, Philip Costanzo, Kerry Haynie, Truls Østbye, Emilio Parrado, Barak Richman, Miller McPherson, and Lynn Smith-Lovin—for their direct and indirect influence on my work and this book.

Three years of my work on the American Civil War were also supported by the Josiah Charles Trent Associate Professorship in Medical Humanities. Mary Duke Biddle Trent Semans has long been a supporter of the history of medicine at Duke, and just about every aspect of my job here has been touched by her generosity. The Trent Collection, which forms the core of the history of medicine library, began with the rare medical books she collected with her first husband, Josiah Trent, and donated after his death. The Trent Foundation bought me the Sanitary Commission microfilms that undergird these pages. Mrs. Semans and the Trent family have supported faculty in the history of medicine at Duke, created two chairs in the medical humanities, and recently established an endowment for the Trent Center for Bioethics, Humanities, and History of Medicine in the medical school. We are all grateful for her support and influence in keeping medical humanities vibrant at Duke.

I also owe a debt to my colleagues in the Duke History Department. Peter English filled in for me during the sabbaticals that made the book possible and has supported my career since he chaired the committee that brought me to Duke in 1992. Other members of the department have offered friendship, information, and the general pleasant environment that makes writing history so much fun. Conversations (not wasted time at all!) with Ed Balleisen, Sy Mauskopf, John Thompson, Alex Roland, Kristen Neuschel, Karin Shapiro, Jan Ewald, Lil Fenn, Peter Wood, Anne Scott, Barry Gaspar, and others have enriched my work. Deans and department chairs—Sally Deutsch, Bill Reddy, Bill Chafe, Karla Holloway, and Susan Roth—have also done much to make this book possible, by saying "yes" when asked for that most precious of scholarly gifts, time.

Time changed radically for me on 4 February 2001, when my son was born. Gone were the weekend days and evening hours for research, reading, and writing. Life on the "second shift" had begun, and my world divided forever into the time before and the time after. Raising this remarkable, intelligent, willful, and exciting child has become the most challenging and rewarding task of my life. It will be many years before he can understand that this is "Will's book," but they have ineluctably grown together.

It has been a pleasure working with Jackie Wehmueller at Johns Hopkins University Press. She too has waited patiently for the "big" Civil War book that the press anticipates, tolerating my diversion into this project. My thanks to her guidance and encouragement, as well as to the anonymous referee who gave the book a timely green light.

Over the past two years I have presented material from this book to audiences at the University of Minnesota, East Carolina University, the College of Physicians of Philadelphia, the University of Virginia, the Waring Historical Library in Charleston, the Society of Civil War Surgeons, the Southern Association for the History of Medicine and Science, and the Congressional Black Caucus Foundation's Annual Meeting. I am grateful for these invitations and very much appreciate the feedback and questions received from these audiences. My favorite presentation was the one in Philadelphia, where black Civil War reenactors in uniform lined the back wall, making the story I had to tell seem real and immediate. I was equally moved by a visit to the African American Civil War Memorial on U Street in Washington, D.C., escorted by the memorial's curator, Frank Smith. I'm grateful for his hospitality and hard work in preserving the memory of the U.S. Colored Troops.

Finally, I come to Ted Kerin, my husband. He has contributed to this project in a thousand large and small ways. He took time off work to cover my usual childcare duties while I traveled to give talks or to visit the archives essential to the project. His computer skills brought out the best in the images that illuminate this work, and he had led me gently into the brave new world of PowerPoint, memory sticks, and mpeg files. He read the entire manuscript, catching typos and suggesting improvements to infelicitous prose. Mostly, though, he has just been there, providing the intelligent, cheerful, patient safe harbor that has anchored my life as wife, mother, and historian.

Intensely Human

CHAPTER ONE

The Black Body at War

It was all going to be over in ninety days. Hurrah, boys, hurrah! But after the battles of Bull Run and Fort Donelson, of Shiloh and Seven Pines, after a year of encounters big and small, the nations faced the reality that this war would be prolonged and bloody. The Confederate government instituted a draft in the spring of 1862 (though a man was exempt if he owned twenty slaves or paid for a substitute). With its larger population, the Union was able to delay taking this unpopular step until March 1863. The monster of war gobbled men at a great rate, and both sides were largely dependent on volunteer armies and popular support to fill them. Once the initial excitement wore off and the casualty lists bore home their grim messages, each side became more desperate to find fresh recruits and to hold on to the men already in the field.

Each side eventually realized, albeit reluctantly, that employing black men in the war effort was critical for success. The South had about 2 million male slaves and used them in all the support functions necessary to military action—building fortifications, cooking, driving wagons, nursing, and repairing infrastructure. By the spring of 1865, the South had even begun enlisting black men into its armies.[1] There were only about 100,000 free black males of army age in the North, but the lands that constituted the Union sphere of influence expanded continually through the war, bringing thousands of black men under the stars and stripes. Northerners were not anxious to make soldiers out of black men, especially those who had recently been slaves, but as the demand for soldiers increasingly drained the white male population, resistance to black enlistments waned.[2] Eventually, black troops would come to represent some 10 percent of the Union fighting force.

These black soldiers never fully overcame the prejudice that accorded them second-class status within the army. Advocates for African American troops struggled to secure them equal pay, adequate provisions, appropriate clothing, intact tents, and working rifles. Nowhere was separate but unequal more pronounced than in the bodily condition of the black soldier. Many black troops entered the war physically handicapped by recent and remote insults to their health, and they suffered more intensively than their white comrades from disease once they joined the ranks. Their experience provided pundits with much fodder for cant about the inherent weakness of the African.

Arming the Black Man

The decision to enlist black men into the Union army was tied up in ambivalent northern attitudes toward abolition and slavery. Lincoln famously said in an 1862 letter to Horace Greeley that his goal was first and foremost to preserve the Union. "If I could save the Union without freeing any slave, I would do it and if I could save it by freeing all the slaves, I would do it and if I could save it by freeing some and leaving others alone, I would also do that."[3] Lincoln drew strength from the abolitionist sentiment, but he had to balance that constituency with the proslavery voters in the border states. During 1862 he increasingly realized that abolition would win the United States moral high ground in Europe and perhaps be key in preventing European alliances with the Confederacy. Finally, after the battle of Antietam lifted Union spirits and stalled an apparent Confederate drive toward victory, Lincoln presented the Emancipation Proclamation. It was an ambivalent document, freeing as it did only those slaves in states under rebellion (over which Lincoln claimed jurisdiction by his war powers) while leaving in bondage black men and women who resided in states still in the Union.[4]

Lincoln and the Union government had been pushed to this decision by the action of slaves who freed themselves by crossing Union lines. At first officers had returned the slaves to their owners, citing the fact that the Fugitive Slave Act was still in force. But Gen. Benjamin Butler, camped with his men at Fortress Monroe on the Yorktown peninsula, would not tolerate such barbarism. He recognized that every slave working on the farms and in the factories of the South freed a white man for service in the army and that without slaves the southern economy would collapse. He accordingly declared the black slaves who had come across his lines to be "contraband of war" and just as appropriate for confiscation as donkeys and chickens captured from the enemy. This policy was approved by the Union government. As Union troops moved through Virginia, Kentucky, Tennessee, Missouri, and Arkansas during 1862, black slaves escaped to them, and army officers found themselves administering growing refugee camps with little experience or supplies. Black men were quickly put to work behind the lines, constructing fortifications, caring for livestock, and other manual labor required to maintain army life. To a lesser extent black women were hired on as cooks, laundresses, and nurses.[5]

It was obvious to some that these willing black men would make strong soldiers, soldiers with a burning commitment to establish their own freedom and the freedom of their brothers still in bondage. In March 1862 Maj. Gen. David Hunter, then commander of Union forces on the South Carolina coast, declared that state's slaves emancipated and ordered the black men into Union army regiments that he created. He neglected to have his actions authorized by his commanders, however, and his actions proved too much, too soon. Similarly, Brig. Gen. Jim Lane organized men in Kansas into black regiments that saw battle before being accepted as legitimate by the Union in January 1863. In Louisiana, affluent free blacks in New Orleans organized their own regiments. They first offered to fight for the Confederacy and only later signed up as Union troops. Initially staffed by black officers, these regiments saw the humiliation of having those men demoted and white officers put in their places.[6]

These initial efforts and talk about expanding black recruiting sparked significant opposition in the North. Arming the black man ran counter to the ruling philosophies of the border states where slavery was defended and justified. Some argued that blacks were cowards and would run from battle. Others claimed that black men could not live as freemen and that even white officers could not keep them healthy, orderly, and fit for battle. Enlisting black men highlighted slavery as the war's cause and emancipation as its main purpose, a rationale highly unpopular with some northerners who were willing to fight to preserve the Union but not to free the slaves. Still, as northern casualties mounted, even the most racist began to see the value of enlisting blacks. Samuel Kirkwood, governor of Iowa told Gen. Henry Halleck in August 1862, "I have but one remark to add, and that in regard to the negroes fighting; it is this. When this war is over and we have summed up the entire loss of life it has imposed on the country I shall not have any regrets if it is found that a part of the dead are niggers and that all are not white men."[7]

Kirkwood may well have been responding to a law passed by Congress three weeks earlier. The law "authorized [the president] to receive into the service of the United States, for the purpose of constructing intrenchments, or performing camp service, or any other labor or any military or naval service for which they may be found competent, persons of African descent, and such persons shall be enrolled and organized under such regulations . . . as the President may prescribe."[8] This convoluted language emphasized the black man's role as a laborer but allowed for his induction as a regular soldier. The law further provided that "if he was a slave to a person in rebellion, then he and his family shall forever after be free." But if he was a slave to a loyal Union man in, say, Kentucky, his family was not liberated. Finally, his pay was set at $10 per month, with $3 taken out for clothing. White privates received $10 per month as well, but they received a clothing allowance of $3 on top of their salary, meaning that white troops made $3 more per month than black troops.[9]

The recruitment of black men got off to a slow start in the fall of 1862 but picked up steam the following spring. Stanton sent Gen. Rufus Saxton to the coasts of North and South Carolina to enroll former slaves as Union soldiers. The governor of Massachusetts recruited blue-blooded abolitionists to lead the state's new 54th and 55th regiments, to be composed of free black men. In the spring Gen. Lorenzo Thomas traveled west to supervise the recruitment of black soldiers in Kentucky, Missouri, and the Midwestern states. Many of the recruits had recently been slaves, or came directly from the plantation. Loyal slave owners received compensation when their men signed on; slaves owned by men with known Confederate sympathies were just taken as contraband of war.[10]

Some plantation owners, loyal or not, resisted the impressment of their slaves into the Union army. They threatened reprisals on the families of men who volunteered to go. The women who remained would have to do the man's full job in the fields, or the women would be sold "down river." Slave women were subject to physical reprisal, be it rape or brutal beatings, if their men left. It is hard to imagine that slave owners living in Kentucky and Missouri and Maryland during the spring of 1863, months after the Emancipation Proclamation had gone into effect, did not recognize that slavery's days were numbered. Surely many were glad to receive any money for this evanescent form of property. But it took an act of Congress (passed in March 1865) to liberate the families of black soldiers whose owners were not in rebellion.[11]

Most of the black men who fought in the Civil War were former slaves and were recruited in Confederate or border states. Information is available about where the men signed up for service, but what is not known is their place of birth or how recently before enlisting they had been in servitude: 88,000 came from states in rebellion; another 46,000 originated from border states where slavery prospered. Of the remaining 25 percent of black soldiers, some proportion had been born in slavery. More than 3,000 men signed up in the District of Columbia alone, and it is likely that many of them started the war in bondage in Virginia or Maryland.[12]

The summer of 1863 dispelled the notion that black men would not fight. Black regiments "saw the elephant" at Fort Wagner outside of Charleston, and at Milliken's Bend and Port Arthur, Louisiana. Charles Dana told William Stanton that "the sentiment in regard to the employment of negro troops has been revolutionized by the bravery of the blacks in the recent Battle of Milliken's Bend. Prominent officers, who used in private to sneer at the idea, are now heartily in favor of it."[13] The heroics of the 54th Massachusetts at Fort Wagner received great play in the northern papers. The black troops grew in public estimation with each battle, and northern sympathy was even more aroused when, in April 1864, Nathan Bedford Forrest's men slaughtered black troops who had surrendered at Fort Pillow in West Tennessee.

Black troops fought in Virginia in 1864 and were there to liberate Richmond in the spring of 1865. They did not, however, march through Georgia with Sherman, who remained skeptical of their fighting power. "I believe the negroes better serve the Army as teamsters, pioneers, and servants," he told Lorenzo Thomas in June 1864. "I must have labor, and a large quantity of it. I confess I would prefer three hundred (300) negroes

THE BATTLE AT MILLIKEN'S BEND.

In this sketch of the battle at Milliken's Bend, black men are portrayed in the same heroic style that *Harper's* artists used week after week to depict white troops. *Harper's Weekly*, 4 July 1863.

armed with spades and axes than a thousand as soldiers."[14] Sherman's attitude prevailed through much of the military hierarchy. Most of the black regiments were stationed away from the front lines. They guarded prisoners, forts on the Mississippi River, and supply lines in conquered territories. They manned ports all along the Atlantic coast and posts around the capital. And everywhere they were assigned, black men were put to heavy fatigue work, just the sort of manual labor that Sherman had in mind when arming his hypothetical soldiers with axes and spades.

The second-class status of the black soldier was brought home to him daily. His tents were rejects returned as worn out by other regiments. His food was bug ridden, lacking in vitamins, and too sparse to keep a man going at heavy labor. His clothing was shoddy, his shoes inferior. Many regiments drilled without rifles, and when weapons finally arrived, they often didn't fire or were much less accurate than the white troops' armament. Medical care for black soldiers was inadequate, and their hospitals were poorly supplied. Even if the white officers told them they were free, life in the army looked far too much like the slavehood days they had supposedly left behind.[15]

For better or for worse, for glory, country, and the liberation of their people, some 180,000 black men wore the brass buttons of the Union soldier's uniform. More than 33,000 men were buried wearing it, with

4,000 of them bearing bullet wounds and the rest defeated by disease. Understanding this excessive mortality from disease requires closer examination of the black bodies that went to war.

The Health of Black Recruits

Historians have long used height as a signifier of nutrition and health. Childhood malnutrition and chronic diseases stunt growth. Conveniently for the historian interested in the health of Civil War recruits, height was measured as part of the recruitment physical. After the war the provost marshal general summarized information gathered during the recruitment examination. He reported that the mean height of black soldiers was 66.21 inches. American-born white males were an inch and a half taller, with a mean height of 67.7. There were substantial differences by region—soldiers who had been slaves west of the Mississippi were, on average, 63.32 inches tall, while coastal state slaves had a mean height of 67.56, near that of white native-born troops.[16]

There is a large literature and much controversy about how well slaves were fed in the American South. In their book *Time on the Cross* (1974), Robert Fogel and Stanley Engerman used calculations of food consumption to argue that slaves had an adequate and nutritious diet.[17] Critics have challenged their figures on various grounds, including whether the excessive caloric needs of hard-working slaves could be met by the diet described.[18] Other historians have looked instead to the height of slaves as a way of indirectly measuring nutritional adequacy. Richard Steckel found that when slaves were transported on ships or trains their height was listed on the shipping manifest, and this led him to discover heights for slave children as well as adults. He reported that the height of slave children was below that seen in the poorest regions of Africa today, suggesting poor diet and chronic disease. By age 7 the children had only reached the first percentile of the height charts used by modern American pediatricians. Not until the teenage years did substantial growth appear. This is consistent with the expected growth spurt of adolescence but probably also indicative of the fact that dietary enrichment occurred when the teenage slave began adult labor. By 18 the male slave averaged 65.3 inches; by 25 he had reached 67.17 inches, slightly taller than the height recorded by the provost marshal for black army recruits. Some of the slaves and free blacks inducted into the army were teenagers who still had some growing to do if Steckel's numbers are correct.[19]

Historian Dora Costa has looked at the data gathered by Benjamin Gould for the Sanitary Commission with questions about height and health in mind. Gould measured men as they left the army, so the men in his sample are by definition survivors. Costa found that black men who were age 16 to 20 at time of measurement were 65.5 inches tall. Black soldiers older than 20 were about 67.5 inches—about the same heights as adult white soldiers (67.6). She points out that the Gould values for adult black men at the end of the war differ so slightly from the averages for black soldiers on admission that taller height (and hence better nutrition prewar) was not selected by the many pathological pressures of army life. In other words, the taller men who likely were better nourished before they went into the army were just as likely to die as those who were shorter and less well fed.[20]

Work by Chulhee Lee on patterns of mortality during the war may illuminate the relevance of taller height to wartime mortality from disease.[21] He contrasted the experience of white troops from rural and urban settings. He found that the rural troops were twice as likely to die of diseases as those from urban settings but that this excess mortality occurred during the first nine months of a soldier's enlistment. This discrepancy existed despite that urban troops were almost an inch shorter on average.[22] Although rural troops contracted illness at a higher rate, their case recovery rate was actually better than their urban comrades. Lee argued that the rural troops grew up in healthier environs and were less likely to be malnourished than their urban peers. Hence they weathered illness better than less well-nourished city dwellers. But they also were less likely to have previously endured illnesses like measles, chicken pox, smallpox, and typhoid, and so were particularly vulnerable to them when first exposed in Civil War camps. Their prior good health could not prevent the onset of illness, even if it did make recovery more likely.

Black troops likely entered the army with the worst possible combination of backgrounds. Available evidence indicates that many would have suffered malnutrition for much of their lives, like white soldiers from, say, the slums of New York. But unlike those urban soldiers, most black troops had come from rural areas and also had been isolated from childhood diseases. They had the rural soldier's proclivity for contracting infectious diseases and the urban soldier's lowered resistance. Smallpox, pneumonia, dysentery, measles, typhoid, and other infectious diseases therefore cut deadly swaths among them.

Information about the health of the men upon entry into the army presents conflicting images of their robustness. One set of data comes from that report published in 1875 by the provost marshal general. After the war he surveyed recruitment station physicians for their opinions about the men they examined. Out of 188 physicians surveyed, 115 responded. It is probably a fair surmise that the physicians who were most conscientious in their examination duties were also the most likely to fill out the survey, but there is no way to verify this assumption. The recruitment center physicians had a fairly high opinion of the black recruits. In fact, 66 percent said that blacks were equal or even superior to white men in their physical qualities. While one in four black recruits was rejected for health reasons, the rate of white rejection was even higher, one in three.[23]

These numbers have several limitations. First, to say that black soldiers were healthier than whites is not to say they were healthy. If the white men presenting to the recruiting stations were a particularly sickly lot, such as the slum dwellers of northern cities were likely to be, then the blacks may have been likewise ill-suited for service but just better in comparison. A second issue is how few recruits these physicians reported seeing. These figures include the examination of just over half a million men, including 26,000 African Americans. Thus only about 11 percent of the total black enlistment is represented in these figures.[24] For whatever reason these soldiers may well have been healthier than the other 89 percent.

Other sources suggest that many black men arrived at barracks in a weakened condition. Ira Russell, the Massachusetts army physician who took a special interest in black patients, reported that "little discrimination was used in the selection of negro soldiers." He summarized his own experience in St. Louis and the recollections of other physicians on the East Coast by reporting that "large numbers in feeble health, with impaired constitutions, broken down by exposure and privation while escaping from their masters, or from over crowding in contraband camps and bad and insufficient diet, were enlisted more with a view of filling up

Ira Russell (1815–88). Photograph used by permission of University of Arkansas, Special Collections Library, Fayetteville, Arkansas.

companies than promoting the efficiency of the service."[25] Comments about the number of men who showed extensive scarring from beatings or had hernias that rendered them unfit for duty were common.

Morbidity and Mortality during the War

The most striking fact about the mortality patterns during the war is that white battle mortality was much higher than for blacks. Around 90,000 (4.5%) white soldiers died in battle or later of wounds sustained against the enemy; a little more than 3,000 (1.8%) blacks endured this fate. A major reason for this disparity was the choice to send black soldiers to garrison posts, assignments where they guarded a fort or other facility behind the lines. This "saved" the supposedly superior white fight-

TABLE I.I
Death Rate from Specific Diseases

Disease	White	Black
Diarrhea and dysentery	17.3 (34,565)*	33.9 (6,108)
Typhoid and typho-malaria	15.6 (31,115)	8.8 (1,581)
Pneumonia	7.4 (14,738)	29.1 (5,233)
Tuberculosis	2.6 (5,286)	6.7 (1,211)
Smallpox	2.4 (4,717)	13.0 (2,341)
Measles	2.1 (4,246)	5.0 (901)
Remittent fever (malaria)	1.9 (3,853)	5.6 (1,002)
Scurvy	.0002 (383)	.002 (388)

Source: *Medical and Surgical History of the War of the Rebellion (1861–1865)*, vol. I, pp. 636–641, 710–12. Hereafter *MSHW*.
*Numbers in parentheses indicate total number of deaths.

ing men for the battlefield. Many blacks did see battle, and not a few died from wounds, but their mortality from disease was far higher. While 2.7 white soldiers died of disease for every single battle casualty, among the black soldiers the ratio was about ten to one. All told, about 13.5 percent of white soldiers died while in the army during the Civil War. Among black soldiers the figure was 18.5 percent. So even though blacks saw much less combat, they died at far higher rates than white soldiers.[26]

The major diseases that killed black and white Civil War soldiers were infectious diarrhea or dysentery, pneumonia, typhoid fever, tuberculosis, measles, and smallpox. Comparing disease-specific mortality rates, however, reveals that blacks were twice as likely as whites to die of diarrheal illnesses but only half as prone to death by typhoid. Pneumonia proved fatal at four times the rate among black soldiers compared to white; scurvy killed the black troops ten times more often. Tuberculosis was three times more deadly among blacks than whites, while the smallpox death rate among blacks was five times that of whites. Even malaria, to which blacks were supposedly less susceptible, killed almost three times more black soldiers than whites.

The 65th USCT Regiment

The worst record for disease and death belonged to the 65th U.S. Colored Infantry Regiment. This group of men had the misfortune to be stationed at two of the deadliest locales of the war, Benton Barracks in St. Louis and the Mississippi River camps in Louisiana. Historian Paul Steiner has chronicled their fate in painful detail, using the "carded records" of each soldier to create a group portrait of pervasive disease and death. Steiner explained, "The regiment was selected for study because of its exceptionally bad health record, because it was organized late in the war when military medicine had advanced about as far as it would go . . . and because it was never in combat." Of the approximately 2,000 Union regiments, the 65th USCI had the most deaths from disease and no combat deaths. Because so many men died (some thirty before they could even be mustered in), the army had to eventually enroll 1,707 men and 62 officers to approach the usual complement of about 1,000 men per regiment. Of these enrollees, 742 enlisted men and 8 officers died of disease or camp injuries. Even by the standards and knowledge of the time, most of these deaths were avoidable.[27]

This regiment not only suffered exceptional mortality but also, not surprisingly, a persistent level of disabling disease. Much of the time only half the men were on their feet, and at times the number available for duty sank below 200. Pneumonia was a major killer in crowded winter barracks; chronic diarrhea plagued the summer months, and these men suffered from scurvy in a land of plenty. Malaria, measles, typhoid, and tuberculosis all took their toll. The regiment did not muster out until 1866, so they were together long enough to be attacked by the third major U.S. cholera epidemic.[28] The morbidity and mortality rates of the 65th USCI regiment may have been extreme but only by matter of degree. Their experience was echoed throughout the ranks of black troops, especially those stationed along the Mississippi River.

The Civil War was a horribly bloody conflict that decimated a generation of young men, North and South. The excessive morbidity and mortality of black troops during the war stood out even amid this vast experience of death and disability. Contemporaries expected the black body to respond differently to war and for the most part found their expectations and prejudices displayed in the medical statistics. Others saw in the numbers a condemnation of the army and its treatment of black soldiers. All agreed that the differential was significant and that it had direct relevance to broader questions about the future of these recently enslaved people and their adaptability to full American citizenship. For a time the black body in health and disease became central to the broader discourse on reinventing the American polity.

CHAPTER TWO

The Pride of True Manhood

The decision to enroll black men into northern armies was made amid a complex discourse about the black body and its capacity for full manhood. Some northern reformers argued that the great transition from slavery to freedom could be partially effected by turning slaves into soldiers, and soldiers into full citizens. "Slave to soldier to man" would solve the most pressing social disruption of nineteenth-century America, the liberation of some 4 million slaves. While this slogan ignored slave women, their status entered this discourse in discussions of the male slave's adherence to the marriage vow and his likelihood of supporting his wife within her proper sphere. Others claimed, however, that the black body was inherently too weak to allow this transition to take place. Southern apologists alleged that the black man was naturally a slave and could not live successfully in any other condition, while a subset of northern abolitionists felt that black men could live on a level of total equality. Yet the vast middle of northern thinkers saw weaknesses in the black body and questioned the degree to which these were biological or subject to the amelioration of education and enlightened attention. The induction of black men into northern armies opened a vast laboratory for investigation and study.

The 1850s and 1860s were times of high anxiety about the black man and his fate. Southerners feared slave revolts and the restriction/destruction of their prized peculiar institution. Northern abolitionist literature increasingly challenged the morality of slavery, especially utilizing the figure of the mulatto to indict the southern male as a fiend who not only sold his own children but took his daughters as concubines. Even as southerners built a defense of slavery on the biological predisposition of the black man for slavery, the presence of mulattoes challenged the rigid distinctions inscribed in such biracial assumptions. Affluent southern women voiced increasingly open disgust at the existence of mixed-race children, while slavery was growing ever whiter. The black and brown body had come to center stage of slavery's defense.[1]

In 1862, when refugee slaves swarmed Union camps and Lincoln penned his Emancipation Proclamation, it became increasingly clear that numerous black people would achieve their freedom and perhaps even flood northern cities like the Irish before them. This prospect raised great anxiety in the north about their ability to live as free men and productive citizens. What was to become of the black man? Would he make a citizen? How was postwar society to evolve? Could he take on the role of true manhood, and would his body allow it? Northern intellectuals were in a flurry to find out. Would all the mulatto men, who appeared to be the intellectual and social leaders of the black world, really die out due to their inherent weakness, or continue to lead their darker brothers toward civilization? All was in flux at midcentury.

Isaac Strain, Officer and Gentleman

In 1865 Joseph Smith, the medical director for Union troops in Arkansas, reported his observations about the health of the black soldiers under his care. After noting that that these troops were obedient and courageous, Smith qualified his praise. "They however have not the intelligence of the white troops and must be made to take proper steps for the police of their persons and their camp. His [*sic*] moral and intellectual culture is deficient and the lack of this culture renders him unequal to the white soldier in power to resist disease." That this deficiency was key to their achieving citizenship—nay, even full manhood—was driven home by his next statement. "That there is a connection between this power to resist disease and death and high mental and moral culture was never more clearly exemplified than in the exploration of the isthmus of Darien by Lieut Strain and party whose adventures and sufferings and endurance have become historical."[2]

So who was this Strain fellow? Why did he exemplify white manhood, much less its power to "resist disease and death?" At thirty-three, Naval

Lieutenant Isaac Strain was one of the most famous Americans of his time. Although efforts in the late nineteenth century to pierce the Isthmus of Panama with canal and rail are well known, the race to find a rapid way across central America began even earlier, with explorers seeking likely routes in the 1850s. In January 1854 Strain led a party of men from Caledonia Bay across the isthmus, seeking a route through the mountains that was low enough for excavation of a canal to be feasible. With the transit to the Pacific only forty miles long at this point, it seemed a simple matter to reach the other side and provide a detailed report on the topography that lay between. Strain was in a great hurry and relied on maps and descriptions generated by others which proved to be wildly inaccurate. Surveyors from France and England were also racing to find a good route and secure the glory and profit for their countries. Strain set out with twenty-seven men, expecting a trek of no more than ten days through jungle rich in game and tropical fruits. He emerged forty-nine days later, nearly dead from starvation and exposure. Strain had split his group, leaving behind the sick and slow while he took a few of the strongest to find help. He returned with a rescue party to deliver the remnant. A third of his party died along the way or in the immediate aftermath of the trek, all to prove that there was no route suitable for a canal in this part of Panama. The mountains were far too high.[3]

From initial inspection it appears that Strain failed miserably. Yet an article in *Harper's Monthly Magazine* downplayed the poor judgment that led Strain and his men into trouble, instead emphasizing Strain's cleverness and heroism in meeting adversity.[4] It celebrated his perseverance in battling his way out of the jungle and leading a party back to find the rest of his men when he himself could barely stand. J. T. Headley, the author of the piece, presented Strain as the ideal officer. "Strain, the leader, though half naked and a small man, was knit together with iron sinews, and with as brave and resolute a heart as ever beat in a human bosom. Fertile in resources, and with that natural spirit of command which begets confidence and insures obedience, no man could be better fitted for the trying position in which he unexpectedly found himself."[5] Such officers were smart, selfless, and devoted to duty.

Strain left his second-in-command to govern those too weak to move when he left to get help. When Strain returned, this officer said "*Did I do right?"* Headley exclaimed of this response, "How that involuntary excla-mation honors him—exalts him above all eulogy! *Duty* had governed him from first to last; *duty* occupied him even in the extreme suffering of star-vation. So long as we can have such officers to command our ships, our navy will retain her old renown, and whether flying or struck, our flag will still be covered with glory."[6] Headley's account frequently extolled the prudence and forethought of the officers and the foolishness of the com-mon men, who rashly gobbled up their rations or damaged key imple-ments, such as the only fishhook. Officers had to guide the men in every aspect of self-governance, just like Smith's black soldiers, who "must be made to take proper steps for the police of their persons and their camp."[7]

But Strain's story did not just reveal the white officer as wiser than his followers and more committed to duty. His moral and mental fiber made him a stronger man, far stronger than those with more apparent brute strength. As the expedition's situation worsened, "a marked and striking difference was seen in the power of endurance between the officers and gentlemen of the party and the common seamen . . . thus proving, what every man has observed who has been in long and trying expeditions, that intellect and culture will overbalance physical strength." It is exactly this distinction that Smith was making about his troops in Arkansas. Headley goes on, "The power of a strong will—the effort demanded by the calm voice of reason and the pride of true manhood—take the place of ex-hausted muscles and sinews, and assert, even under the pangs of famine and the slow sinking of overtasked nature, the supremacy of mind over matter, of the soul over animal life, no matter how vigorous the latter may otherwise be." These officers not only remained strong, able to battle ad-versity as it came, but even when met with disaster, they "were as active, energetic, and cheerful as at first."[8]

Historian Stephen Berry has described a southern concept of success-ful manhood as the acquisition of éclat, a worldview that included flam-boyance, ambition, and the courage boldly to enter uncharted lands car-rying flag and civilization.[9] Strain was the very model of the modern man and officer. His plans to rush across Panama were undeniably bold. While he initially tried to treat the indigenous peoples with fairness, he soon found them treacherous and decidedly in need of civilizing influences. Strain felt no compunction about his imperial schemes. He had permis-

sion from the governments that nominally ruled the area and never considered that the Indians who opposed his passage were merely defending
their own lands. Strain embodied manifest destiny and carried American
imperialism unabashedly into the jungle. If the jungle proved it would not
be so easily conquered, Strain nonetheless became the hero of the hour,
symbolizing all that was great, good, and strong about the American way.
Even five years of civil war had not quelled in Smith's mind this glim-
mering image of manhood's ideal.

When the black soldier suffered from disease or fatigue, white officers
far too readily called upon such rhetoric to condemn the black body as
lacking in *endurance*. It is hard for a modern historian to grasp how anyone who was familiar with the daily grind of plantation field work could
question the capacity of the black man to endure hardship, but it was common to do so. Perhaps because black superiority in bodily strength threatened the accepted hegemony of white over black in all things, those ques-
tioning the endurance of the black body coupled endurance with vitality,
with mental activity, with moral strength. The *Harper's* article specifically
contrasted the endurance of puny but mentally superior Strain over the
burly seamen of his party, which included one black man. Likewise the
muscular black field hand somehow had to be weaker than the diminu-
tive white college boy, if contemporary perceptions of the rank of races
were to be preserved.

In response to a U.S. Sanitary Commission (USSC) survey that asked
about the endurance of black soldiers and how their "physical conforma-
tion" affected military performance, one physician scoffed at the possibility that blacks had the same capacity for endurance of white men.[10]
"Certainly not. Taken in a body, they are, *at present*, too much 'animal' to
have moral courage or endurance." For example, he explained, "The negro [is] quite as amenable to the treatment of diarrhea as the white man,
except that he is deficient in moral courage and endurance."[11] Another
agreed that the black soldier lacked the physical endurance of whites, for
"he is more easily depressed in spirits when attacked by acute disease."[12]
The survey also asked about the endurance of mixed-race men, a topic
considered in detail later in this chapter. One might expect that the combination of supposed traits, mental superiority from the white ancestry

and physical strength from the black, would meld into a being superbly fitted for endurance. But such amalgams could not pass as superior given contemporary rhetoric about racial inferiority. Hence, said one physician, "The experience of every medical officer I have found whose views were at all noted on the subject is that while an infusion of white blood tends to mental activity and vigor, it is at the expense of physical power and endurance."[13] One Massachusetts physician feared that "any considerable admixture of white blood deteriorates the physique, impairs the powers of endurance, and almost always introduces a scrofulous taint."[14]

There were some who pointed out the absurdity of this argument about lessened endurance. A physician at a recruiting station in Troy, New York, found great powers of endurance in the men he examined. "The colored man [sic], as far as my observation goes, make excellent soldiers. They are, as a race, remarkably free of hernia, are muscular, and capable of great endurance."[15] A physician carrying out similar duties in New Hampshire felt the black man's endurance was obvious. "Does not the infamous and cruel history of the race sufficiently attest it?"[16] Yet the observations of these physicians did not actually go very far, for they saw the men only at the beginning of their service, not after time in the field. In Alabama one officer described his black troops as remarkably strong. "When parched corn, beef and half rations of salt composed their diet scarce a word or murmur was heard. When we consider all these things, we see endurance was not wanting. They are men possessed not only with a cheerful and willing spirit—easily handled, but have also corresponding physical strength."[17] Even this praise has a grudging tone, with the caveat "when we consider all these things." It could not be denied that the black soldiers were unfortunately prone to disease, and some justification had to be offered. The argument that they lacked mental fortitude and, hence, bodily endurance, made for a handy explanation that avoided the implication that they had been treated poorly by the army's bureaucracy. Only the white soldier, and especially the white officer, could embody the total manhood so evident in Lieutenant Strain.

"O My Brethren! Are We MEN?"

The equation of full citizenship with manhood runs through nineteenth-century rhetoric on race relations. "A freeman in a political sense, is a citizen of unrestricted rights in the state," wrote black nationalist Martin Delany in 1852. Black men in the United States had been "shorn of their strength, disarmed of manhood, and stripped of every right," by the actions of white people, but Delany determined to remedy the loss with education about black accomplishment.[18] The exclusion from full participation in the polity rested on the assumption that black males lacked the qualities of true manhood. Black reformers struggled against this judgment. "Are we MEN!!—I ask you, O my Brethren! are we MEN?" exclaimed David Walker in an appeal to his fellow "men of color" in 1829. As historian Mia Bay has argued, "Walker's question . . . would resonate throughout nineteenth-century black writing on race," a literature that, "like its white counterpart, was written almost exclusively by men and aimed to find a place for black men among the 'races of men.'"[19]

Much of the language that spoke of the black male's elevation was written by well-meaning white authors, and their voices predominate in this account. Delany argued that black people must elevate themselves but acknowledged there were many white people who mistakenly thought it was their duty to perform this function. "The coloured people are not yet known, even to their most professed friends among the white Americans," Delany explained. This familiarity, or the lack of it, was a constant variable in rhetoric about the black body. "Politicians, religionists, colonizationists, and abolitionists have each and all, at different times, presumed to *think* for, dictate to, and *know* better what suited colored people, than they knew for themselves."[20] And they would keep doing so in the coming decade, as the status of black people came to the forefront of American political rhetoric.[21]

Black authors and white seemed to think that the transition from slavery to freedom (or elevation from poverty for the free black) required a transition to full manhood. But if the black adult male was not fully a man, then what was he? There were various answers to this question, and each answer had implications for whether the black body would support trans-

formation. One common assumption was that the black male was a being somewhere between animal and full human, perhaps closer to an ape
than a human being. A second trope visualized the black male as a perpetual child, with the white placed in a paternal position over him. Yet another line of thought aligned the black male with woman, sharing with
her qualities of dependence, love of beauty and arts, religious fervor, and
frivolousness. A final image depicted some northern free blacks as full
men but the southern slave as a degenerated creature, a being formed in
response to climate and poor treatment, who could be elevated given
proper education and intervention. Not surprisingly, all but the final description came predominantly from white authors.

The analogy of black men to beasts was pervasive in nineteenth-
century American culture. This was particularly the case for the black slave,
as the U.S. Constitution itself established that they counted, after all, as
only three-fifths human in the census-based calculation for congressional

CUTTING HIS OLD ASSOCIATES.
Man of Color. "Ugh! Get out. I ain't one ob you no more. *I'se a Man, I is!*"

This cartoon illustrates how common the association of black people with animals was in American society—and shows that this assumption was now being
questioned, at least in a northern periodical. *Harper's Weekly,* 17 January 1863.

representation. Slaves were put up on the auction block like horses and pigs. Potential buyers probed their bodies as they would a dumb animal's to discover age, recalcitrance, and strength, and families were separated with their owner's giving them no more thought than a child handing out kittens on the town square. Women were advertised as "good breeding stock," and slaves were lashed like poky mules to work them harder. Slaves were listed in wills right along with the rest of the livestock. If the chattel slave was to become a man, he had somehow to escape this brutish lot.[22]

In her analysis of former slave reminiscences gathered by Works Progress Administration interviewers in the 1930s, Mia Bay describes the overwhelming presence of animal analogies in accounts of slavery days. Again and again these people used animal metaphors to express the misery of subordination under slavery. They remembered sleeping on the floor like hogs, eating out of troughs like (and sometimes with) pigs, going barefooted as a duck, being worked like a mule, or getting beaten like a dog. Just like animals, they lived in drafty, dirty sheds, were not allowed in the main house, and were assumed to have no feelings. In contrast, if accorded kindness, former slaves said they had been treated just like white folks. To be black was to be like an animal; to be human was to be white.[23]

The analogy of blacks to animals fit smoothly into proslavery arguments. In the two decades preceding the Civil War, apologists had moved from considering slavery as a necessary evil to portraying it as a positive good.[24] Their argument turned on the assumption that black people were incapable of achieving mature adulthood and only under the benevolent guidance of the white owner would they prosper and become civilized. This inferiority was inbred, and some authors went so far as to claim that blacks were not fully human at all and were not descendants of Adam and Eve. Rather, there had been a separate Creation, overlooked by biblical writers, that generated the black-skinned natives of Africa. Mobile physician Josiah Clark Nott was the most prominent proponent of this theory of polygenesis, which allowed slavery's advocates to even more securely relegate the black man to the animal world. No amount of education could, after all, make a chimpanzee into a human. Black people were similarly constrained by their bodies to a lower caste position. This view was not widely adopted, even in the South, for it flew in the face of biblical literalism, but that the idea received any hearing at all reflected the power of the black-man-as-beast idea.[25]

Such depictions highlighted the black man's supposed lack of higher mental functioning—he is dumb, numb, and lacking in sensibility. But another version of the bestiality metaphor presented the black man as oversexed and lacking in the moral control that would prevent unbridled expression of his sexuality. Before the Civil War such hypersexuality was more often attributed to black women, in part as an apology for white male sexual predation in the slave quarters.[26] Fear of the black male as a bestial rapist of white women would become far more common in the Jim Crow South than before the Civil War, when the potential for general slave uprisings aroused more panic than the possibility of black ravishment of white women. Still, it was present. In the few instances where black soldiers were accused of accosting white women during the war, the incidents received wide publicity in the southern press.[27]

It was not uncommon in the mid-nineteenth century for writers interested in the natural world to compare human beings to apes. That these creatures resembled people was obvious even before Darwin, and ethnologists used apelike features to suggest that Africans were intermediate between the ape and white man. Researchers calculated facial angles, measured the distance between fingertip and knee, or assessed the shape of the heel, all in an effort to show the supposed apishness of the black man. One writer sympathetic with the elevation of the black race lamented that "this low, degraded and miserable section of the genus homo, this people whose claims to humanity have been so often peremptorily denied" should be "classed with Apes and Monkeys and Baboons."[28] Orangutans were familiar animals in urban curiosity shows, and their sexual appetites were reported on with salaciousness. Historian Elise Lemire draws the connection between titillating representations of women being carried off into the jungle to be ravished by orangutans and cartoonish images of black men with exaggerated apelike features pawing white women. She even postulates that when Edgar Allan Poe revealed that the culprit in his story "The Murders in the Rue Morgue" was an orangutan who had climbed in the window, a reference to the predatory black male would have been clear to his readers. She claims that this point was further made

by the murder weapon. The ape used a barber's razor, and the story postulates that far from meaning to cut the female victim's throat, the ape
thought he was only mimicking a barber. In 1849, 5 percent of the black
work force of Philadelphia were barbers.[29]

Whether the black man was a predatory ape or a docile, dumb brute,
such bestial metaphors made it clear that his body constrained social development. The orangutan or chimpanzee could be dressed up and taught
some tricks but could never be made into a man. Likewise, if the inferiority of black people were biologically determined, then advocates for raising the black male to full manhood were doomed to failure. This argument existed in all sorts of variations. At one extreme stood the polygenist
who believed the black man was as different from the white as the ape
from the human. But more common were those who accepted some degree of difference and assumed it was biologically based. Chapter 3 ex-
plores the many bodily variations assigned to the black man, such as
differential susceptibility to disease, climatic affects on health, and endurance of suffering. Remaining at question was whether the black body's
differences were so marked as to preclude its equality with whites within
the polity.

Defenders of slavery also liked to depict the black male as a child,
fondly known as Sambo. "Sambo possessed all the virtues and vices of the
white child," explains historian Joel Williamson. "Like a child he could
know love and loyalty to the parent. He was a physical person, often com-
ical, energetic, playful, fun-loving, and innocent. But like a child he could
also be careless, improvident, animal, and thoughtlessly cruel." The
tropes of animal and child could clearly cross. "Without white guidance
he would destroy not only others but himself as well. With enlightened
white guidance, he would survive, be made useful, and flourish."[30] This
conception allowed the white slaveholder to see himself as an amiable father to a vast family, providing only the loving discipline that any good parent bestows on his child. Unmentioned within this beatific description of
the plantation household was that some of those slaves actually were his
children, unacknowledged and vulnerable to all of slavery's indignities.

The "slave as child" metaphor provided obvious advantages for southern rhetoricians. It painted the southern white slaveholder as benevolent
and also allowed him to say to outsiders, "you don't know these people as

well as we do. We have their best interests at heart. And if you don't think children need discipline, well then you've never raised any." It fit into nineteenth-century notions of what children owed their parents and how households were ideally ordered. "I am nominally your slave until I am of age," one southern son could say to his father, "but I glory in having you as my master."[31] Yet unless child-ness was linked with some inherent biological basis, the argument raised a problem for the defender of slavery. Children grow up and respond to education and discipline by becoming full-fledged adults. Their bodies are inherently mutable. If the black man was indeed a child, he need only be educated properly to join the white in full manhood.

The contrasting images of man and animal or man and child, however despicable, grew rather obviously out of the situations of slaves on plantations. Another comparison, the black male as female, makes less immediate sense. In many ways the stereotypic traits of the black man and white woman were polar opposites. The white woman was viewed as delicate, refined, unable to bear hard labor, and in need of protection. Nothing could be further from the image of the brutish black male. Yet in many legal discussions in the North, the free black and the white woman shared features of dependency and legal inferiority. Historian Rowland Berthoff has looked at the published debates of twenty-four constitutional conventions held in northern and western states during the antebellum period. There he finds much anxiety expressed among the white male speakers over the potential domination of white females and black males, and the concomitant determination to deny them the vote, keep them off juries, and limit their rights to hold property. Like women, black men could not be trusted to make clear decisions, avoid demagoguery, and manage financial affairs. Their freedom was accordingly limited, and the full rights of manhood denied them.[32]

Black people supposedly displayed other characteristics that allied them more with women than mature, self-reliant white men. Officers writing about their black troops were not hesitant to comment on such effeminate qualities. According to stereotype, black soldiers loved bright clothes and wasted what little money they had on decorative gewgaws to show off to their fellows. This love of flash made the soldier's uniform with its shiny buttons a real draw at recruitment fairs. The race had a natural love of music and dancing, a trait that served them well on the parade ground where soldiers learned to march and drill to the beat of a drum. Observers of black religious services found men and women lost in the ecstasy of the spirit, overcome with joy through the preaching and the music. Such softness and susceptibility was ascribed to the feminine congregant; she was the opposite of the mature white male who wore somber clothes, enjoyed restrained music and disciplined dance, and kept a calm visage at Sunday services. It was all about control, and blacks and women were clearly wanting in such self-discipline.[33]

Mixed with these various stereotypic depictions of the black male was a language that spoke of *degradation*. Degradation (or degeneration) implied that a person had slipped from some prior state of rectitude and thus might yet be returned to it.[34] Southerners were particularly prone to view urban free blacks as degenerate, seeing these undisciplined people as prone to all sorts of vices, especially that of attracting slaves into misbehavior and even rebellion.[35] Others referred to northern free blacks in similar terms, ones that would have been fairly interchangeable with discussions of all slum dwellers, including the immigrant Irish. They were lazy, prone to drink, sexually debauched, and thieving.[36]

This language about degeneracy could also be used by those sympathetic to the plight of slaves. Frederick Douglass wrote in 1854 about the degeneration of the black body. "The form of the Negro has often been the subject of remark," he told an audience at Western Reserve College. "His flat feet, long arms, high cheek bones, and retreating forehead, are especially dwelt upon, to his disparagement."[37] Rather than seeing these bodily changes as permanent, however, Douglass felt they were the result of cruel treatment and poor climate. Reversal of such injustice would return the black man to his noble form and allow full ascent to manhood. Later, Douglass would famously declare that a uniform could effect that transformation. "Once let the black man get upon his person the brass letters 'US,' let him get an eagle on his button and a musket on his shoulder and bullets in his pockets," said Douglass in an August 1863 manifesto, which called on Philadelphia's black men to enlist, "and there is no power on earth which can deny that he has earned the right to citizenship in the United States."[38] He urged his listeners not to be dissuaded by the unequal treatment in pay and assignment so far bestowed on the black soldier by the Union government. They must seize the opportunity to liberate their race, for "the opportunity is given to us to be men."[39]

Slave to Soldier to Citizen

"The problem is solved. The negro is a man, a *soldier,* a hero." So proclaimed a white officer impressed by the performance of his black regiment in combat near Petersburg, Virginia, early in the summer of 1864.[40] Another officer saw the change occur even before his black troops experienced the elevating fire of battle. The commander of the 59th U.S. Colored Infantry described the induction process for freedmen. The men stripped off their old clothes, tossed them in a fire, and stepped into a bath. Then they buttoned up their new blue coats, and the metamorphosis was complete. The officer enthused, "Yesterday a filthy repulsive 'nigger,' to-day a neatly attired man; yesterday a slave, to-day a freeman; yesterday a civilian, today a soldier."[41] Many white officers of black regiments were less sanguine that the problem would be so quickly resolved but had confidence that making the black male a soldier, would, ultimately, bring him to manhood.

This rhetoric sought to alleviate deep national anxieties about the place of black people in the American polity, especially if the 4 million slaves in bondage at the start of the war acquired their freedom. President Lincoln had issued the Emancipation Proclamation in late September 1862 (although it did not go into effect until 1 January 1863). It seemed increasingly likely as the year drew to a close that liberation would come to these benighted people, bringing massive and threatening change to American society. A Missouri newspaper editor reflected these fears in November 1863, commenting on a recent gathering at Leavenworth to promote black equality. "It is impossible to look upon these circumstances and indications without alarm. Unless the new wave of fanaticism is thrown back, it will bid fair to overwhelm the land with the most momentous consequences," he warned. This author had no faith at all in the transformative power of the black regiments. " 'The slave of yesterday, the soldier of to-day, the citizen of to-morrow,' is a phrase, the adoption of which as a matter of true doctrine must involve the country in the destructions of republicanism's essence, and give it over completely to the sway of ignorance and demagoguery." How could it possibly work that the ignorant black man, "taken from the cotton field, clothed in blue and brass," was to become an "enfranchised American citizen, the social and political peer of the FILLMORES and the BANCROFTS, and EVERETTS and SEWARDS of the land." The editor concluded, "We have no stomach for the new gospel."[42]

Even those more sympathetic to the plight of the freedman feared the "momentous consequences" of liberating 4 million people who had never lived as independent citizens. After Benjamin Butler refused to return runaway slaves in 1862, and declared such people to be "contraband of war," the northern government scrambled to establish a refugee policy and find ways for dealing with large numbers of destitute ex-slaves who had escaped to Union lines. In response, Secretary of War Edwin Stanton created the American Freedmen's Inquiry Commission to investigate "the condition and management of emancipated refugees." Appointed 16 March 1863, the commission members traveled to various southern venues under Union control and questioned many witnesses about the social behavior of the former slaves. Several concerns were paramount. Would these people work for a living if not forced to it by overseers? Would they create stable families in which husbands and wives cohabited and jointly raised their children? Could they provide self-care when sick and call upon physicians when necessary? Could they be taught basic reading, writing, and calculating skills? One commission member traveled to Canada to find out if the black communities in Ontario were thriving and particularly whether their bodies could remain healthy in the cold climate.[43]

Not surprisingly, the commission often received conflicting testimony, but a few clear conclusions emerged. Their final report did contain a long and scathing denunciation of slavery and all its cruelties, prodding the Congress to create the Freedmen's Bureau in partial response and to ease the transition to freedom. The Commissioners found that most witnesses reported the contrabands were happy to work or enlist in the army, as long as they were treated fairly. Illiteracy was widespread and the rudiments of elementary education badly needed. Many witnesses attested to the weakness of the "marriage bond" and family allegiances. Slave marriages had not been legally sanctioned, and the adult male slave had never been able fully to assume the husband's role, being unable to protect his wife and children from abuse and separation. Widespread disease and death in refugee camps called into question black families' ability to care for each other and preserve health. The report recommended the establishment of freedmen's hospitals, which did become one facet of the Freedmen's Bureau operations. Its final sections endorsed the enlistment of former slaves as soldiers, and called for the complete emancipation of the slaves, with no interim period of oversight or limited freedom.[44]

"Abolitionist officers were the most ardent advocates of using the army as a school for citizenship education," argues historian Keith Wilson. This idea was not limited to those who were totally committed to black equality. "There was throughout the officer corps a general and all pervasive belief in the benefits of army life. Billy Yank, black or white, had to learn appropriate patterns of behavior; he had to learn the connection among power, authority, and service."[45] Army officers sought to instill order, discipline, honor, and, at times, concepts of hygiene. But the black soldier, and especially the recently enslaved, required more. Education was the most obvious need, as so many of the enlisted black men could neither read nor write. One officer wrote about his black troops camped near Vicksburg, that "Church will only be held when in his opinion it will be for their best interests, as education [it] is believed, should be the paramount object in elevating these soldiers to the true status of manhood."[46] If there was time left over from military duties, it should be spent learning.

Other efforts on the part of the white officers of black regiments reflected the drive to elevate the soldier's status. When officers recognized that their men engaged in religious services characterized by undue enthusiasm and even witchcraft, they offered instead the more staid ceremonies of mainstream Protestantism. Officers were entitled to legally marry men and women in areas under Union control, and some made as much of the ceremony as possible, to reinforce the importance of the marriage rite and the participants' responsibility for continuing within it. Some Union officers felt an obligation to act as role models and mentors for their men, to somehow instill in these recruits habits of industry, self-discipline, and civilized morality. The army was, in the word of Thomas Higginson, colonel to a regiment formed from liberated slaves in the Sea Islands of South Carolina, to be their "university."[47]

But becoming a soldier was a *bridge* to manhood, not the achievement of manhood itself. Although some aspects of being a soldier marked the acme of manhood—courage, patriotism, devotion to duty—others did not fit the stereotype of the free male citizen. The soldier had to follow orders and was not free to come and go as he pleased. The penalties for desertion included death; at least the captured runaway slave could expect his expensive life to be spared. This may have been particularly difficult for the southern white male soldier, who was not used to being ordered about by other men. "'It grinds me to think that I am *compelled* to stay here,' Joshua Callaway [C.S.A.] remarked typically. 'I've got a dozen masters, who order me about like a negro.'"[48] Sam Watkins of the first Tennessee regiment likewise found loss of manhood in the soldier's life. He had volunteered for twelve months, only to see his obligation extended to the end of the war without his having any choice about it. "When we were drawn up in line of battle, a detail of one-tenth of the army was placed in our rear to shoot us down if we ran. No pack of hounds under a master's lash, or body of penitentiary convicts were ever under greater surveillance." Watkins then made the most negative analogy of all. "We were tenfold worse than slaves; our morale was a thing of the past; the glory of war and the pride of manhood had been sacrificed upon Bragg's tyrannical holocaust."[49]

Watkins reported that errant Confederate soldiers were whipped as punishment, highlighting even further the slave-soldier similarity. Screenwriters for the movie *Glory* had the soldier played by Denzel Washington flogged for going AWOL, allowing for the dramatic moment when the regiment's colonel sees that the soldier's naked back bears witness to prior beatings under slavery. In reality, officers of black regiments realized that the correlation of whipping with slave punishments sent the wrong message to black volunteers who sought glory, honor, and manhood through military service. Although they allowed a number of other cruel practices, commanders expressly forbade punishment by whipping and court-martialed officers who disobeyed the order.[50] As a category of punishment, whipping had held special meaning as far back as ancient Rome. There, part of full male citizenship was freedom from whipping, for beating was punishment that was not only painful but demeaning to manhood. Slaves, women, and children were all vulnerable to this bodily penetration, but men were not. Soldiers fell in a special category—they could be whipped, but only with a special vine staff that was the centurion's mark of office. In the Civil War the soldier likewise occupied a liminal zone between full manhood and slave. The black northern soldier was protected from the lash to preserve his transition to manhood, while the white southern soldier endured the punishment so commonly used to degrade the vulnerable within his own culture.[51]

There were other characteristics of the stereotypic soldier that made his situation less than ideal as a bridge to manhood. After admitting that his black troops "seem the world's perpetual children, docile, gay, and lovable, in the midst of this war for freedom on which they have intelligently entered," one officer mused that "it seems to be the theory of all military usages, in fact, that soldiers are to be treated like children."[52] The military promoted discipline, nay, even mindless discipline. The soldier had less to think for himself than to do as he was told. Drilled into him was the concept that he was merely a part of the machine, an interchangeable cog of the larger mechanisms that were his company, his regiment, his corps. The soldier had to give up individual will and individual self-discipline to the discipline of the group. Finally, there was irony in the assumption that giving a black male a gun and training him to violence would make a man of him. Somehow the scripted violence of the battlefield created manhood while the untamed violence of the slave revolutionary indicated only bestial savagery. Underlying all the rhetoric, though, was the idea that white men were ultimately in control and would bring order to the black soul, order that the black soldier would somehow internalize and carry forward to a productive life as a new citizen.

But could the black body endure the rigors of the soldiers' life? Was the transformative treatment too harsh for the child/animal/woman to bear? Some observers were quite ready to find their endurance wanting, while others defended their stamina. Yet it was in resistance to disease that the black soldiers seemed to fare worse than their white comrades. The mortality rates in contraband camps were appallingly high, and regiments of black troops reported similarly grim statistics. This apparent weakness threatened to undermine the slave to soldier to citizen mantra so prominent in northern paeans about the process of inducting the former slave into the army.

Examining the Brown Body

The existence of light-colored African Americans vastly confused this rhetoric about the black body and its limitations. Historians have long recognized how important the mulatto was in debates about slavery and the freedman. If whiteness and blackness were not distinct categories, and instead every sort of skin tone existed in America from blue-black to café au lait to pale peach, then arguments about bodily limitations lost their precision. The existence of the light-skinned slave embarrassed southern politicians who wanted to brand the northern abolitionist as promoting "amalgamation," the label given to the sexual mixing of whites and blacks on an equal and legal basis. Clearly, the southern slaveholder had no problem countenancing amalgamation when it occurred on his own terms. In writing novels with an antislavery message, authors favored the "tragic octoroon," a nearly white girl threatened with sale into prostitution or, in one memorable instance, pursued by her own father for his sexual pleasure. When southerners claimed that northern factory owners treated their workers worse than slaves, northerners could reply that at least they did not sell their own children for profit.[53]

The mulatto drew increasing attention as the war drew nearer. The newspapers highlighted court cases in which slaves claimed their freedom on the grounds that they were really white, and in Virginia a person could acquire a legal judgment affirming that he or she was "not black."[54] In the federal censuses of 1850 and 1860, data gatherers for the first time counted mulattoes as well as black and white residents. Although at times mulatto had been a label for a specific degree of black blood, in the mid-nineteenth-century United States it was an imprecisely defined term implying that the observer recognized both Caucasian and African traits in the person before him. No definition was supplied to the census taker; it was assumed he could judge the matter by sight. In 1850 enumerators found 247,000 slaves who were mulattoes (7.7% of all slaves); by 1860 that number had grown to 412,000 (10.4%). As historian Joel Williamson has noted, slavery was getting whiter and whiter. In parts of the west, such as Missouri and Kentucky, one slave in five showed evidence of white ancestry.[55]

If, as southerners wanted to argue, the black man was innately inferior and intended by God for his servile position, then what to say about the mulatto? It was blatantly obvious, as Charles Sumner happily pointed out, that "the best blood of Virginia flows in the veins of the slaves."[56] Some southerners tried to dismiss the problem by downplaying the numbers of mulattoes. Others pointed to the mulatto as the best hope for the black race and saw the mixing as a form of racial elevation. Many were no doubt relieved when Josiah Clark Nott, who had put forward the polygenic argument so effectively, pronounced the mulatto a hybrid, like the mule. The mule was infertile, and likewise the mulatto had low fecundity. Any such people generated by black-white sexual liaisons would be sickly and die out in a few generations. The mulatto was, in other words, a temporary result of racial mixing but by no means an obstacle to the biological basis of slavery. This rationalization allowed slavery's apologists to ignore the mulatto, which they did explicitly, or imply that any degree of black blood fitted a man to slavery, which they did implicitly.[57]

At the start of the Civil War, northern physicians had many questions about the black body, but little firsthand experience with black patients. There was widespread familiarity with Nott's arguments about hybrids and interest in exploring these questions, as exposure to light brown bodies allowed. When the U.S. Sanitary Commission created its survey to explore the "Physiological Status of the negro," it included two questions relevant to the mulatto: "8. How do the mixed races compare to with pure negroes?" and "16. Effects of amalgamation on the vital endurance and vigor of the offspring."[58] It appears from the Sanitary Commission archives that these surveys were mailed to individual physicians during the war and were carried about to various sites where black patients clustered and administered to physicians by USSC investigators.

It may be that USSC interest in the mixed-race soldier was heightened by the fact that he was overrepresented among U.S. Colored Regiments. It was commonly believed that men of color with evident white ancestry were more intelligent than their darker brethren, and when recruiters chose from a population, lighter-skinned men may have been preferred. The census of 1860 listed about a third of the free "Negro" male population in the United States as mulatto, and many men from this group volunteered as soldiers. Union recruiters drew heavily from plantations in

Maryland, Kentucky, Missouri, and Arkansas, freeing the slaves chosen and enrolling them in the army. These were areas where 10 to 20 percent of the slaves appeared as mulatto in the census. Finally, autopsy registers in St. Louis from 1864 include the race of the man on the table; 75 percent were noted to have some admixture of white blood. The health and endurance of the light-brown body may have been of particular interest to those studying the troops of color.[59]

While many answers to the survey's questions appear in the Sanitary Commission's archives, there is no summary document that attempts to cumulate the results. Assumptions that mulattos were smarter but somewhat weaker were fairly common. Other respondents noted no difference at all. The most in-depth account of the question was written by Ira Russell, who traveled down the East Coast in the summer of 1865 studying a variety of issues relating to the health of black soldiers. He had earlier directed those autopsies in St. Louis and found that men of mixed race appeared healthier than the darker men. "In the colored regiments recruited at Benton Barracks, Mo., the mixed race largely exceeded the pure blacks," he reported. "They were more robust, and, on average, about an inch taller. The average height of *100* mulattoes was *5* feet *8* in., that of *100* blacks, *5* feet *7* in."[60] Historians looking at other Civil War data sets have found similar differences in height between men labeled pure black and mulatto, lending credence to Russell's findings. The height difference may indicate that mixed-race slaves and freedmen suffered less malnutrition in childhood than their darker peers.[61]

Russell's research generally tended to debunk the accepted argument about the bodily weakness of mulattoes. In July 1865 he spoke with Dr. J. D. Harris in Richmond, a black physician who had treated many of the U.S. Colored troops during the preceding year. Harris was "very emphatic and decided in his opinion that the admixture of the races, does not impair physical endurance or fecundity, but on the contrary, promotes both." Russell goes on to contrast this opinion with that of Dr. [Joseph?] Jones, Rebel and ex-slaveholder, who said that the "black race will become extinct by being absorbed in the white race and that this absorption will be beneficial to both." Russell found Jones's evidence for this point rather extraordinary. "To prove his position . . . he called in two of his ex-slaves, both

mulattos and both bearing a very striking resemblance to himself and ex-
hibited them as specimens of physical culture."[62]

Russell was not satisfied in limiting his inquiries to doctors. He figured
that if anyone knew whether men of mixed race were puny and weak, or
women of mixed race did not succeed in producing offspring, it would be
the slave trader. So off he went to talk to the men who had run Richmond's
biggest slave markets, "to learn what effect the admixture of white blood
had upon the market value of the slave." Russell was told that for field
hands, there was no perceived difference by skin tone, but for "the farm
hands, hotel waiters, house servants, mechanics, etc. an admixture of
white blood enhanced their value, other things being equal." The difference was most marked for lighter-colored women. "Female slaves were
often purchased, to use the language of these traders, for 'fancy use' and
valued in proportion to the preponderance of white blood and the possession of personal beauty." Russell concluded, "in deciding this question
of physical endurance, I place great confidence in the instincts and observation of the Slave Trader. While such men as De Bow have held the
theory that amalgamation tends to deterioration and extinction, the more
sagacious and practical slave dealers, were very careful not to let any such
notions interfere with their practice or profit." All in all, "the conclusion
would seem probable that mulattoes do perpetuate freely."[63]

After the war, interest in the mulatto faded as the end of slavery
brought an end to the figure's usefulness as a pivotal point in slavery arguments. If before the war the free people of mixed race in southern cities
served somewhat as a middle caste between white and black slave, after
the war they merged with the 4 million freedmen to become, simply, people of color. Increasingly, either people were "white enough" to pass or
they were black, determined by any evidence of African heritage. Pundits
later in the century predicted the eventual demise of the entire "inferior
race" without applying niceties of skin tone to the prognostication.

As American intellectuals came to realize that black emancipation
truly was imminent, they responded with a confused discourse about
manhood and citizenship. Existing literature about black distinctiveness
had largely been written either to defend slavery or to promote abolition.

THE SLAVE OWNER'S SPECTRE.

And the Nigger never flitting, still is sitting, still is sitting
On that horrid bust of HORACE just above my chamber-door;
And his lips, they have the snigger, of a worthless freeborn Nigger,
And he swells his sombre figger, when I ask him, with a roar,
"Will you blacks again be Cattle, as you used to be before?"
                          Cries the Chattel, "Never more!"

This parody of Edgar Allan Poe's "The Raven" emphasizes the slave owner's
despair that the docile black slave shall be seen "nevermore." *Harper's Weekly,*
30 May 1863.

Even scientific works, supposedly grounded in objective evidence, tended to be framed in political language. Most northerners knew little about black people, and this was particularly true of northern physicians. Accordingly, white discourse about the black body and its potential took place in an atmosphere divorced from any experience of the thing in question. This was not true of southerners, by and large, but their agenda in the defense of slavery left northern thinkers unsure about whether any such observations were valid.

The fascination with the mulatto offers one example of this multilayered discourse. Nott's pronouncements about the weakness of mixed-race men and women was accepted fairly widely in the north, as after all he was a southern physician and should know. Yet Russell could find few southern physicians, or slave dealers for that matter, who had seen any evidence of the argument's validity. Mixed-race slaves worked well, reproduced well, and were valued as house servants and concubines. The more northern physicians were exposed to black men of various shades, the more they dismissed the argument as groundless. Claims about the peculiar health status of mixed-race persons fit within the ideology of slavery's defense and disappeared with the peculiar institution itself.

The mantra of slave to soldier to citizen offered reassurance to northern society that this vast body of liberated men could indeed be civilized and controlled. If the black male was an animal, or a child, or even a woman, it made sense that discipline and education would transform him into an acceptable member of society. The taming of the beast, the boy, and the shrew were familiar tales. Still, the soldier was not quite a man in full, and there appeared little hope that the black man would ever be, either. It was possible to transform the black male out of childhood, out of bestiality, and away from the feminine, but his mental weakness, proclaimed by most every commentator on the black body, prevented entrance into full manhood. Black soldiers could not be officers; they could not attain the height of manhood embodied in Lieutenant Strain. Without the necessary mental and moral capacity, the black soldier could never achieve the *endurance* that led Strain to conquer Panama. Within this discourse, the black man could achieve, at most, a sort of junior manhood that entitled him only to the most basic perquisites of society.

CHAPTER THREE

Biology and Destiny

Physicians and other learned men of mid-nineteenth-century America were certain that the black body was biologically different and distinct from the white one.[1] This conclusion went beyond general assumptions about manhood and endurance to embrace a catalog of various conditions of difference. So widespread was this perception that Abraham Lincoln stated it as a fact in 1858. "There is a physical difference between the white and black races which I believe will forever forbid the two races living together on terms of social and political equality."[2] He was addressing the (to him) unlikely possibility that blacks and whites would ever routinely marry or otherwise cohabitate, an outcome that Stephen Douglas had thrown up as the inevitable result of ending slavery and allowing even minimal civil rights for African Americans. Lincoln was drawing on a common assumption that blacks and whites were physically different in significant ways. Physicians of the time period could have elaborated for him the many ways that blacks varied medically. Their assumptions would be tested and, at times, verified as black men endured so much disease under the observing gaze of northern physicians in Union hospitals.

This chapter has several goals. One is to review the modern debate over racial disparities in health care in order to craft an approach to historical questions of racial disparities. The second is to review nineteenth-century conclusions about racial differences in disease susceptibility and response to treatment. The third is to try to use modern medical research, where applicable, to shed light on the phenomena described by nineteenth-century writers. Such an approach does privilege twenty-first-century western medical knowledge, but to my sights it is the best resource on the subject now available.

Debating the Biology of Race

The first challenge for the medical researcher seeking to understand how racial and ethnic groups experience disease is the problem of defining who, exactly, is a member of such groups. Since human beings can breed across racial and ethnic lines without difficulty, any particular person can claim more than one race, defying attempts to neatly categorize him or her. As a position paper for the American Anthropological Association has argued, "It has become clear that human populations are not unambiguous, clearly demarcated, biologically distinct groups . . . Throughout history whenever different groups have come into contact, they have interbred." It is this very pattern of commingling that has maintained humankind as a single species. Thus, "any attempt to establish lines of division among biological populations is both arbitrary and subjective."[3] David Brion Davis agrees, noting that since science offers no precise definition of racial groups, "the so-called races of mankind are the fortuitous and arbitrary inventions of European and American history, the by-products, primarily, of Europe's religious, economic, and imperial expansion across the seas of the earth."[4] Others go so far as to claim that employing any category of race as if it has biological reality is to yield to the pervasive racism of our culture.[5]

These authors, and it is fair to say that theirs is the most visible point of view in today's academy, call for the elimination of racial categories because they have been so abused in the past. The most egregious example concerns the use of race to discuss differences of intelligence, the variation in IQ scores by race. Stephen Jay Gould's book *The Mismeasure of Man* is iconic in exposing the ways in which supposedly objective science has been employed to reinforce concepts of racism.[6] Herrnstein and Murray's *The Bell Curve*, with its analysis of IQ by racial categories, epitomizes the evils that follow if race is reified.[7] Not only is the acceptance of race as a biological category fraught with the danger of racist abuse, goes this argument, it also is based on very weak science. Humans all interbreed;

we are one species. There is no single gene or marker that determines whether one is white or black or Pacific Islander or whatever. How can you do a study on a biological variable when it cannot be defined? The exercise is ludicrous.

Yet it is very hard to abandon racial categories. We have fields, departments, and careers built on the study of African Americans. These scholars all seem to know who they are talking about and would have difficulty in redefining or renaming their fields. Americans are quite comfortable with stating their race on the myriad forms that appear before us, and only a few choose the new category of multiracial or more than one race. Historian Barbara Fields admits as much. "The rubric of the hour is race. Though discredited by reputable biologists and geneticists, race has enjoyed a renaissance among historians, sociologists, and literary scholars. They find the concept attractive, or in any case hard to dispense with, and have therefore striven mightily, though in vain, to find a basis for it in something other than racism . . . The effort to redefine race as culture or identity is bound to come a cropper just as did the effort to define race as biology."[8]

Fields refers to the fact that historians have tried to come up with other terms, such as difference with a capital *D,* to properly discuss race in this current environment. Fields wants to emphasize that racial labels are about imposing an identity on the person so stamped against his or her will. She mentions a tragic case in which police officers shot a black man during an encounter, only to find that he was an off-duty policeman engaged in no crime whatsoever. She suspects that the shooting victim probably put his identity as a police officer first and a black man somewhere later; his attackers gave race the first priority. Fields considers scientific and medical researchers to be backward in their use of the concept of race, yet even she accepts the use of racial categories when it comes to affirmative action. Her logic emphasizes that this acceptance is about righting wrongs, curing racism, and avoiding stereotypes rather than endorsing race as a fixed category.

Indeed, those same medical and scientific researchers are likewise struggling to figure out the right way to talk about population diversity. The U.S. government, which funds so much of medical and scientific research, requires that researchers take race and gender into account when designing the study population. Here is the guideline on race: "The 1997 OMB revised minimum standards include two ethnic categories (Hispanic or Latino, and Not Hispanic or Latino) and five racial categories (American Indian or Alaska Native, Asian, Black or African American, Native Hawaiian or Other Pacific Islander, and White)." The National Institutes of Health (NIH) emphasize that researchers must collect information about these categories and requires grant applicants to design their clinical trials with sensitivity to them. But the guidelines go on to make it clear that these divisions are social, not medical or biological. "The categories in this classification are social-political constructs and should not be interpreted as being anthropological in nature. Using self-reporting or self-identification to collect an individual's data on ethnicity and race, investigators should use two separate questions with ethnicity information collected first followed by the option to select more than one racial designation."[9]

The government is walking a fine line here. The Tuskegee syphilis study made clear the vulnerability of African Americans to abuse at the hands of scientists. So the government wants to be careful that such minorities are treated appropriately. On the other hand researchers have tended to do most research on white males, with unclear relevance of their findings for females and non-Caucasian racial groups. So the NIH wants to be sure that such populations are not excluded. How is the researcher to define racial categories? The NIH says ask the subject—that is, use self reporting.

A 2002 editorial in the *Raleigh News and Observer* heralded a new program at the University of North Carolina at Chapel Hill called Ethnicity, Culture and Health Outreach. Its goal is to teach health professional students about racial disparities in health care and research, and to aid in developing inclusive protocols. The lead to the editorial reads: "Generally speaking, disease doesn't discriminate by race or gender. Some medical professionals do, unfortunately, when they fail to include adequate numbers of African-Americans or women in research studies."[10] In other words, we must acknowledge race in order to prevent discrimination.

Likewise, a major Institute of Medicine report calls for new research initiatives on racial and ethnic disparities in health care: "Racial and ethnic minorities tend to receive a lower quality of healthcare than non-minorities, even when access-related factors, such as patients' insurance status and income, are controlled . . . A comprehensive, multi-level strategy is needed to eliminate these disparities."[11] Even this report, which acknowledges the existence of racial groups in order to redress inequities in health care, is unsure how to deal with the reification of racial categories. "Some researchers have speculated that biologically based racial differences in clinical presentation or response to treatment may justify racial differences in the type and intensity of care provided [as in use of drugs like enalapril] . . . These differences in response to drug therapy, however, are not due to 'race' per se but can be traced to differences in the distribution of polymorphic traits between population groups."[12]

Other authors want to replace race as a category with discourse that uses the language of population genetics. In 2003 Robert Schwartz proclaimed in an editorial in the widely read and influential *New England Journal of Medicine,* "Race is a social construct, not a scientific classification . . . The degree of multiracial identification [revealed in the 2000 U.S. Census] underscores the heterogeneity of the U.S. population and the futility of using race as a biological marker." In spite of this, Schwartz noted that race-based medical research is common. "Such research mistakenly assumes an inherent biologic difference between black-skinned and white-skinned people. It falls into error by attributing a complex physiological or clinical phenomenon to arbitrary aspects of external appearance." After discussing the concentration of sickle hemoglobin among people of African descent, Schwartz argued, "They reflect geographic origins, not race . . . This is not to deny that the frequencies of certain allelic variants or mutant genes among people who share a geographic origin or culture have medical value. Obviously, a screening program to detect sickle hemoglobin should focus on populations of African descent, and screening for Tay-Sachs disease in New York should be confined to Ashkenazi Jews."[13]

It is hard to see clear guidelines in this sort of statement. It is those arbitrary aspects of external appearance that create the identity that reveals the population history. Medical researchers are in a serious quandary. They want to do the right thing and incorporate black people in their studies as ordered by the NIH. They worry about black people being discriminated against by doctors or hospitals and want to do a study to help correct this wrong. Yet they are being told that there is no way to scientifically define their study population. This flies in the face of common sense, and most just ignore it. People know whether they are black or white or Chinese, and how they self-report is what matters. Others have struggled to bring the debate back to the genome. The premier volume on human diversity defined by genetic geography is L. Luca Cavalli-Sforza's *The History and Geography of Human Genes*. Cavalli-Sforza and his colleagues want to discard the word *race* and replace it with other terms. "From a scientific point of view, the concept of race has failed to obtain any consensus; none is likely, given the gradual variation in existence . . . By means of painstaking multivariate analysis, we can identify 'clusters' of populations and order them in a hierarchy that we believe represents the history of fissions in the expansion of the whole world of anatomically modern humans."[14] So although Cavalli-Sforza and colleagues abandon the term *race*, they bring in clusters of populations or populations based on ancestry for peoples in diaspora communities.

This approach begins to bring some compromise into this discussion. Neil Risch and his colleagues in genetics at Stanford go further, specifically attacking Schwartz and calling for a more reasoned recognition of human groupings. "A major discussion has arisen recently regarding optimal strategies for categorizing humans, especially in the United States, for the purpose of biomedical research, both etiologic and pharmaceutical. Clearly it is important to know whether particular individuals within the population are more susceptible to particular diseases or most likely to benefit from certain therapeutic interventions." Then Risch and colleagues go on to attack Schwartz for the quality of his science. "In our view, much of this discussion does not derive from an objective scientific perspective . . . We demonstrate here that from both objective and scientific (genetic and epidemiologic) perspectives there is great validity in racial/ethnic self-categorizations, both from the research and public policy point of view."[15]

Risch talks about how genetic markers tend to cluster in ways that overlap the geographic location or ancestry of the person tested. Via genetic markers he identifies five major groups—African, Caucasian, Asian, Pacific-Islander, and Native American. Of course, the categories are fuzzy; of course, there is overlap. But it makes sense to suppose that if some gene frequencies cluster in these groups, others might as well. In any given study one cannot predict, however, whether a particular disease pattern is due to inborn differences or environmental factors. That is to be discovered, not assumed. Forensic pathologists work with a list of traits that are more common among certain racial groups than others. One trait does not make the identification, but if enough are present they are able to say with a high degree of accuracy that the skeleton before them was African or Asian or Caucasoid when the skin was still attached.[16] There is no single marker, but the presence of a cluster of traits or a cluster of genes indicates the likelihood (not the certainty) that the person's ancestors predominately originated in a certain part of the globe.

Modern discussions about racial disparities in disease and health care tend to concentrate on access to the lifesaving medicines and technologies that most affect life after the age of forty-five.[17] Our focus here is mainly on diseases that kill children or young adults. If such diseases inflict high mortality rates on a population they have the opportunity to act as strong selective agents, for those who die young will fail to reproduce. Any variation that by happenstance increases survival in the face of such diseases will be selected and preserved. So the length of time a given population has "known" a disease matters in determining how mortal a particular epidemic of that disease will be. These simple facts of evolutionary biology are borne out in the grand scale of world history events, such as the "American Holocaust" that followed the introduction of novel viral diseases to the New World by European invaders from the fifteenth century on.[18]

This mindset is valuable for approaching nineteenth-century thinking on race and medicine. There is no doubt that physicians worked within a highly racist culture, one that could seriously question whether African Americans were even fully human, much less deserving of equal rights and treatment. That they were racist does not rigidly determine, however, that any observations about racial variation in disease must be the result of viewing the black body through a racist lens. It is also possible that black men, and black soldiers, did contract certain diseases at different rates than whites or responded to those diseases in different ways. Some of these differences may have been entirely constructed out of preexisting prejudice or assumptions, some may have been due to genetic differences, some may have resulted from the interaction of environment and genome, and some may have been due to environment alone. At times their black skin was mainly a marker for prior living conditions (slavery, urban poverty), while at others it pointed to the geographic origin of their ancestors in lands with a distinctive, geographically determined set of selective influences. The historian cannot determine this causative mix with anything like certainty, but it is worth exploring the question with an open mind.

Malaria

If antebellum physicians were sure about any aspect of the black man's health, it was that he tolerated malaria much better than whites. This certain knowledge grew out of experiences in colonial South Carolina and Georgia, where, after 1680 severe malaria made the summers deadly for white settlers, while black slaves appeared not to be as endangered. Slavery's apologists touted this difference as one sign that black people were particularly designed by God to live and work in the tropics. When federal troops moved to occupy the coastal regions of South Carolina or to maintain forts along the southern reaches of the Mississippi River, they relearned the lesson that white men from areas where the severe form of malaria was not indigenous were quickly flattened by chills and fever. So one major justification for recruiting black soldiers was that they not only could take over fatigue work but also could be sent to the highly malarious areas of the South, where they would presumably thrive in the environment that was so dangerous to whites.[19]

Resistance to malaria was part of a broader pattern of tolerance supposedly inherent to the black body. In the minds of contemporaries, malaria was tied to a swampy, hot, humid southern place, and black people were thought to function much better than the whites as workers within this steamy climate. After reporting on the good health of black soldiers in Port Royal, South Carolina (and pointing to the many white soldiers who were in the hospital), an African American newspaper crowed, "This is a most striking evidence of the power of the 'American of African Descent' to stand a climate which almost at once prostrates Northern white men, and consigns them to early graves."[20]

To the surprise of many contemporaries, black troops suffered extensively from malarial diseases during the war. According to the official record, there were 2.7 cases per white soldier, and the case fatality rate was 1.73 percent. Among blacks there were 2.4 cases per soldier, but case mortality reached 3 percent. If the figures for typho-malarial fever are added to these, the number of total cases increases slightly, but the mortality rate increases markedly, to 2.6 percent in whites and 5.0 percent in blacks. Even the official history admits that cases likely were underreported. After considering the statistics in detail, its author concluded that high black illness and mortality rates were due to the fact that the black men were "aggregated in the malarious districts" so the excess was entirely due to exposure. He felt that it was still likely that given equal exposure, blacks would do better than whites in a malarious atmosphere.[21]

Other observers recognized that black troops appeared to be just as susceptible to malaria as whites but blamed it on the geographical source of the black men. In his survey of black hospitals in Virginia, Ira Russell found no evidence of protection from malaria and its ravages. He reported instead that only local blacks thrived in the dangerous environments of the James River peninsula. "I am assured that the Planters in this region seldom purchased slaves from the western and mountainous portions of the state [Virginia], for the reason that such slaves, as they expressed it, 'would not live a month['] transferred from a non malarial to a malarial district."[22]

In retrospect it is fairly clear why these events occurred. The black men, women, and children forcibly transported from the West African coast to South Carolina during and after the 1680s brought *falciparum* malaria with them in their bloodstreams. They had grown up in areas where this parasite was endemic and had survived long enough to be kidnapped and sold into slavery. That very survival indicated just that—the ability to survive and function in an environment laden with *falciparum* malaria. In the past fifty years scientists have identified a number of genetic traits involving the hemoglobin molecule and other blood components that protect the bearer from *falciparum* malaria. These traits are common in the population from which many American slaves originated. Probably more important was the complex acquired immunity that each body had achieved by fighting the parasite to at least a truce. That immu-

The Guard of the 107th Infantry Regiment, U.S. Colored Troops, stand before
the guard house at Fort Corcoran, Washington, D.C., 1864. Men like these,
drawn from areas around Washington, had likely never been exposed to *falciparum* malaria. Library of Congress; original negative has crack that has been
corrected electronically here.

nity was maintained by continued encounters with the parasite; it faded
if the person left an environment where *falciparum* thrived. Slave populations in South Carolina remained dense enough to continue the chain
of malaria infection, and the swampy lowlands provided sufficient *anoph-
eles* mosquitoes to allow transmission. "Virgin" whites, those with no prior
experience of *falciparum,* who came in contact with these populations of
slaves would have been highly susceptible to disease and death from this
form of malaria.[23]

Colonial observers had no way of separating out the influence of genetic heritage versus acquired immunity. They also may have been so
overwhelmed by the impact of the disease on whites that they failed to notice that the black slaves did continue to contract malaria, especially the
small children, and to die of it. In the early twentieth century, the death

rate for *falciparum* malaria was twice as high for blacks as for whites in these regions of the South. The reason for this apparent paradox in perception is that blacks had no choice as to where they lived in slavery times and enjoyed less mobility than whites well into the twentieth century. In contrast, whites could choose where to live. In the colonial era most chose more northern colonies; in the twentieth century whites could often afford at least marginally better housing located further from mosquito-breeding swamps.[24]

There are several reasons why African American soldiers in the Civil War surprised contemporary observers with their susceptibility to malaria. First, many of the recruits who ended up in the malarious zones came from areas where *falciparum* malaria was not prevalent. Men from Kentucky, much of Tennessee and Missouri, Maryland, and all of the non-slave states would not have had the opportunity to build acquired immunity to malaria before they were abruptly stationed in an environment teeming with *falciparum*. Second, the incorrect idea that they did not need quinine deprived many of the one life-saving drug in the physician's arsenal. Third, the greater concentration of black men in these malarious regions meant that those men who did have partial immunity could serve as potent malaria carriers to their comrades who were less protected. A white soldier would be struck down by *falciparum* and be confined to his tent or hospital. But a black man with tolerance could have parasites in his bloodstream and be up walking around, allowing mosquitoes to more effectively spread the disease.

Yellow Fever

Antebellum southern physicians were convinced that African Americans suffered less from yellow fever than did whites. They were also aware that once a person had this wretched disease and survived, he or she acquired lifelong immunity. There was much talk in the early years of the war that yellow fever would serve the South as a scourge of the Yankees. After all, in the 1850s, epidemic after epidemic had wracked the Irish and other immigrants to New Orleans, while the region's natives, who had already undergone their "seasoning," were immune. Uniforms would not protect those Yankee bodies when they invaded the Southland. Union commanders were aware of this threat, and one reason offered for sending black troops to Louisiana and the South Carolina coast, both familiar haunts of yellow fever, was to avoid the disease.

As it turned out, yellow fever was not a major presence during the war. In April 1862 Commodore David Farragut overcame the defenses of New Orleans and handed rule of the city to General Benjamin Butler's Union troops. Well aware that yellow fever had visited the city almost every year for the previous two decades, Butler instituted rigid quarantine and put men to work cleaning the streets. Only a few cases appeared in the city while it was under Union occupation, although there were dozens on ships anchored on the Mississippi.[25] In North Carolina, yellow fever erupted in New Bern, Beaufort, and Wilmington but did not extend beyond those cities. There were also outbreaks in Key West and on the gulf coast of Texas.[26] Strangely, yellow fever did figure in an unlikely plot to murder President Lincoln. Convinced that the disease was transmitted through the foul excreta of the yellow fever patient, Dr. Luke P. Blackburn sent a suitcase of linen soiled with such fluid to Washington, D.C. His hope was to get someone in the White House to open it, sparking an epidemic that would claim the president. But the man charged with transporting the infected shirts lost his nerve, and the poisonous valise was never delivered.[27]

The immunity of black troops to yellow fever was thus never significantly tested during the war. There were 190 black cases of yellow fever and 23 deaths reported, all among men in the Atlantic region. In contrast, there were 1,181 white cases and 409 deaths.[28] This mortality differential (12% vs. 35% case mortality rate) supported the presuppositions of contemporary physicians but did not receive any comment in the literature evaluating the black man's experience of wartime disease.

Modern historians have disagreed about whether African Americans indeed carry any inborn protection from yellow fever. As the disease has a long history in Africa, particularly the regions from which American slaves came, it would not be surprising if such a deadly disease had exerted selective pressure on any trait that increased survival. But no such trait has been identified. Historian Kenneth Kiple pointed to the evidence of multiple yellow fever epidemics in the eighteenth and nineteenth centuries that show minimal black mortality. While he admitted that some of this protection was due to immunity acquired through mild childhood cases, he believed that some sort of inherited protective trait was also likely. Historian Sheldon Watts countered that the history can be entirely explained by acquired immunities. While Kiple's evidence is more convincing, the question will not be settled until, and if, modern researchers identify some protective trait that is unequivocally genetic.[29]

Many of the black troops in the Atlantic region, from which the cases and deaths of yellow fever recorded in the official records of the war came, would have been drawn from the slave populations of the area. So it is indeed possible that they had prior exposure during previous epidemics. In contrast, it is unlikely that any of the white, Union troops would have had that experience. Thus the mortality differential may be entirely due to prior immunological experience.

Pulmonary Disease

Just as nineteenth-century physicians had widely held notions that blacks were resistant to malaria and yellow fever, they were equally sure that African Americans were more susceptible to pulmonary diseases such as pneumonia and tuberculosis. "All surgeons any way cognizant of the diseases peculiar to the negro, are unanimous that pneumonia is the bane of the race," reported a U.S. Sanitary Commission surgeon who surveyed medical opinion in the spring of 1865.[30] The war provided mixed evidence for these preconceptions. Everywhere the black soldier was hospitalized he proved particularly vulnerable to pneumonia, and his mortality statistics bear this out. Tuberculosis, on the other hand, was not a major disease among the black troops.

One physician viewing black soldiers suffering from lung disease in a Nashville hospital wrote of the severe cases of pneumonia among them. "Pneumonias attack readily and very generally prove fatal," he reported. "The course of diseases generally, and of those of the lungs particularly is very rapid, terminating frequently in an extraordinary amount of destruction of tissue, entire lung . . . [being] melted down into pus."[31] This susceptibility was attributed to various aspects of the black man's environment. He could not tolerate cold as well and responded particularly badly to the cold, wet winters of 1864 and 1865. One USSC physician concluded, "With lungs naturally tender and a predisposition to tubercle when taken from his cabin where he has been accustomed to a fire the whole year round, and put into a tent with out a fire or into damp barracks, he takes cold and either Pneumonia or Phthisis are induced."[32] In the surveys that the USSC sent to physicians there was near universal agreement that the black soldier was particularly susceptible to pneumonia and prone to die of it once contracted.

The reports on tuberculosis were more mixed. Among black soldiers from the northeast, stationed largely on the East Coast, tuberculosis was relatively common. It was much less prevalent in the West. Ira Russell found that many men died with tuberculosis, which was evident on autopsy, but the disease was not the primary cause of death.[33] In spite of the expectation that blacks were riddled with scrofula (a lymphatic form of tuberculosis), the disease was a relatively minor wartime problem for white and black. There were 3,859 black cases of tuberculous diseases listed in the official record, followed by 1,296 deaths, for a case mortality rate of 34 percent. In contrast, there were 19,890 white cases and 5,418 deaths (27% case mortality rate). These numbers mean that of all the white men that served, about 0.2 percent contracted tuberculosis and died, while the number for blacks was more like 0.7 percent. It is likely that most cases of active tuberculosis were picked up on initial examination or were fairly quickly recognized once the recruit began to endure the hardships of army life. Such men were thus weeded out, either at the recruiting station or shortly after being mustered in.[34]

There is evidence of racial differences in lung volumes that might affect how ably African Americans fight off pneumonia. The forced vital capacity, a measure of lung volume, of black Americans today is about 6 percent lower than whites. This difference was noted by Benjamin Gould, who measured lung capacity among soldiers leaving the army during the Civil War. Researchers had worked with various versions of the spirometer in the decades before the war, but Gould's study marked the first systematic examination of black Americans. He found that "full blooded" black soldiers blew ten to twenty cubic inches of air less than their white comrades. Modern pulmonologists continue to adjust the normal values of spirometry testing by race.[35]

It is difficult to assess whether these differences in pulmonary volume translate into any modern advantage or disadvantage with regard to pneumonia. Likelihood of acquiring pneumonia is related to access to quality medical care. A visit to a health care provider early in the course of an infection may result in an antibiotic prescription that prevents the progression to pneumonia. Regular medical care may include the provision of pneumococcal vaccine, another preventive measure. Underlying conditions, such at HIV/AIDS, renal failure, cigarette smoking, or prior stroke, may increase the likelihood of pneumonia and be differentially distributed across racial groups. Not surprisingly, then, available studies provide mixed results about racial susceptibility to pneumonia. In Ohio researchers recently found that that among males the incidence of pneumonia was 55 percent higher in blacks than whites, while subsequent case mortality was higher for whites (13%) than blacks (9.7%). But when the analysis was limited to men age 18 to 44, the age range of Civil War soldiers, the incidence of pneumonia was doubled and case mortality rate tripled among black men compared to white. This may well reflect the influence of HIV in lowering resistance to pneumonia in this age range.[36]

An assessment of deaths due to pneumococcal disease in California during the 1990s found African Americans to have elevated rates compared to whites. Another hospital-based study in California found blacks and whites to have similar mortality, while Asian and Hispanic patients had better outcomes. On the other hand, a study of 147 Veterans Administration hospitals in the mid-1990s found that blacks had lower mortality rates for pneumonia compared to other racial and ethnic groups. There is no clear message from modern medical science to help explain the Civil War patterns of pneumonia mortality.[37]

Evidence concerning differential susceptibility to tuberculosis is stronger. In the late 1980s researchers from the Arkansas Department of Health examined tuberculosis outbreaks in nursing homes and prisons. Since so much of the debate about racial disparities in tuberculosis has centered on whether the differences are due entirely to environment (availability of nutritious food, degree of overcrowding, access to medical care, exposure to occupational dusts), these environments were particularly valuable because all of these variables were the same across the study population. They found that when an exposure occurred, in this case when a patient or inmate developed active tuberculosis and spread the disease through coughing for the duration that the disease went undetected, blacks were twice as likely to develop clinical disease as whites.[38] This suggests that as a group they are less able to fight off infection once they are exposed. Similar disease susceptibility has been noted among many groups of native Americans. There is some evidence that the difference may be due to macrophage function. Macrophages are cells that engulf and kill bacteria, acting as one of the main lines of defense against the tuberculosis bacillus. Ongoing research seeks to identify genetic markers for what is in all likelihood a complex genetically determined difference in macrophage function.[39]

Others continue to doubt that the differences in tuberculosis rates are due to anything but inequalities in socioeconomic status. Particularly in the modern United States, they argue, poverty can generate many factors that increase the likelihood of tuberculosis. Poorer people tend to crowd together into housing, so that one person with an active case can more easily infect others. The impoverished may have lower access to health care providers and lead the sort of chaotic lives that makes taking pills every day for months on end too complex to handle. These people need help, and focusing on the biology of race (blaming the victim for being black, say) is just a way to avoid that responsibility.[40] Supporters of genomic exploration counter that if genetic differences can be identified, and if their function can be sorted out, it might open new opportunities for medical intervention.

In any event, the pattern of tuberculosis cases in the Civil War makes sense given what we know of exposure rates and place of origin of the soldiers. Physicians who treated black men in northern urban areas likely saw many cases of tuberculosis, as it was a common disease in that setting, and black men would have had ample opportunity for exposure. But most of the black soldiers did not come from urban areas, and so there would have been few tuberculosis carriers to light the fire of an epidemic in the black regiments. There were places, such as in St. Louis in the winter of 1864–65, where tuberculosis did appear and spread quickly among the black troops. But that outbreak ended when the men were dispersed from the crowded conditions of Benton Barracks to multiple sites under canvas in the field.[41]

Flat Feet

The problems associated with flat feet have plagued armies for centuries, for men who cannot walk, stand guard, or march are not of much use in the military.[42] When in May 1865 physicians who staffed the recruiting stations were questioned about their "experience as to the physical qualifications of the colored race for military service," one surprising finding was mentioned by multiple respondents. The African American candidates commonly had flat feet. One physician went into great detail about why "the African race, as a class, are, by reason of the peculiar formation of their bodies, less adapted than other races of men for infantry duty." After claiming that the black pelvis was "light and narrow," the lower extremities "slender, lean and elongated," and the calves poorly developed, he said that such men were not suited for long marches. "The excessive flatness of their feet, (the ordinary arch which exists in the European being almost entirely wanting,) in addition, disqualify them for this branch of the service." Many others mentioned the flat foot or the "splay foot" of the African American recruit. Some were astute enough to note its presence but recognize that "the flat feet do not disqualify for long marches."[43]

These observations, that flat feet were common among black recruits and that they seemed to suffer no ill effects from this "deformity," both have a basis in fact. It turns out that "most flat feet are due to normal biological variability and are considered physiological or flexible." The arches of children are commonly flat, and in some people the arch rises by adulthood. It is only the risen and *then* fallen arch that causes a stiff foot and pain on walking.[44] But variation in the height of the normal adult foot is common. West Africans tend to have a flatter foot as adults. Europeans, on the contrary, have a higher arch. Adults in whom the higher arch never developed have no particular susceptibility to foot pain or malfunction.[45] In fact, the higher arch makes its bearer more vulnerable to damage and disability. Physicians unfamiliar with any feet but those of European descent had to learn these facts as the war progressed.

Other Conditions

Some physicians were sure the black man was particularly susceptible to a host of other conditions. In 1863 the U.S. Sanitary Commission survey on the health of the black soldier sought information on the "causes [that] induce the negroes' incapacity to resist disease" and aimed to find out "under what diseases does he most hopelessly sink?" The assumption that the black soldier was inherently weaker than the white pervades the questionnaire. The form particularly asked about "scorbutic," or scurvy. It also asked about "nostalgia," a form of severe homesickness that incapacitated the soldier with melancholy. There were also questions about therapy, whether what suited the white soldier also benefited the black.[46]

The answers to the questionnaire are scattered through the USSC papers of the Medical Department. Some are mere slips, where someone has cut the answer on, say, scurvy, out of the original document to collect it with others, so the author of the answers is often obscure. In any event, the answers about scurvy, nostalgia, and therapy form no clear consensus. Some said blacks were particularly susceptible to scurvy and needed extra vegetables; other said they were better at scavenging plants from the woods to head off this condition. Some pointed out that the black soldier coming from the plantation had little to be nostalgic about; others said he was particularly prone to melancholy. On therapy there was likewise no agreement. He could not bear "lowering," reported some correspondents, referring to treatments thought to depress the system, like bloodletting and dosing with calomel. Others answered that the black man responded just the same as the white if the doctor paid attention and did his job well.[47]

There was significant discussion of diet in the responses to the Sanitary Commission. The typical southern slave diet consisted of pork as the main meat and corn meal as the main starch. The Union army ration, on the other hand, offered beef and wheat products. Physicians frequently commented on the necessity of supplying ex-slaves a familiar diet. "Accustomed on the plantation to succulent vegetables and fruits he becomes constipated when deprived of them, and this constipation leads to diarrhea of an ulcerative character," wrote one physician to the Sanitary Commission. "On hard tack bacon & coffee he becomes lethargic and weak, and complains of head-ache which is referable to indigestion."[48] Others emphasized the distinction between cornmeal and pork versus hardtack and beef, and pointed out that the former was not only better suited to the black soldier but cheaper too. However, the regiments of free blacks recruited in the North were not happy when this regimen was applied to them. One Union officer stationed in Louisiana reported that his men wanted beef and wheat bread just like the white troops.[49]

The general pattern of presupposition and conclusion about black soldiers and disease which emerged from the war reinforced the general idea that black men were weaker than whites and more prone to disease. Those infections for which he may have had some advantage of increased resistance, such as malaria and yellow fever, were reported in ways that belied his strength. The fact that he may have survived bouts of malaria more commonly than the white man was buried in the overall impression that he suffered so much of it during the war. Even though he did better than whites when stricken with yellow fever, this discrepancy was not remarked on by writers during or after the war. Rather, it was the black soldier's high mortality from all diseases, and pulmonary diseases in particular, that attracted the most attention.

Analyzing infectious disease disparities during the war in light of modern epidemiological knowledge makes some of these patterns clearer. Even if they had higher case survival rates, African American soldiers endured far higher rates of exposure to mosquito bites and malaria, with resulting high rates of disease and death. The world's modern malaria problem is, after all, at its worst in Africa, among African peoples. There were few black soldiers stationed in areas where yellow fever broke out during the Civil War, and their survival differential does not seem to have been of much interest to contemporaries. Without exposure, stark disparities could not appear. It is not clear why contemporaries repeatedly reported that the black soldier was more susceptible to pneumonia; if it was an accurate observation, the reasons underlying the disparity remain unknown. There is clear modern evidence for differences in tuberculosis incidence and virulence by race and ethnicity—but in the Civil War tuberculosis never became a major threat.

CHAPTER FOUR

Medical Care

Even by the standards of the time, African American regiments received decidedly second-class medical care. Like the wag's complaint that the food was terrible and there was not enough of it, for these men there were too few doctors, and many of them lacked compassion and skill. In the best of circumstances, physicians served as vocal patient advocates, protesting inadequate supplies, granting sick leave to weakened men, calling on the U.S. Sanitary Commission for extra food and clothing, and rigidly enforcing contemporary standards of hospital cleanliness and wound care. The mediocre physician ascribed high mortality rates to differences inherent in the black body and spirit, and drank the medicinal whiskey stores while minimally serving his black patients. At his worst, the surgeon was cruel, capricious, and dismissive of the needs of the men under his supposed care. In the hospitals run by surgeons of the latter stamp, the patients received inadequate food and endured filthy conditions. While nurses sometimes took up the cause of the African American hospitalized patient, most of the caregivers tending black men were too weak socially to effect needed reforms. Although it is difficult to quantify the impact of inadequate medical care on the high rates of disease and death among black regiments, the evidence indicates that poor care was an important factor in these outcomes.

Recruiting Doctors for the Black Regiments

Few black regiments ever had their full complement of medical attendants. Official policy dictated that each regiment should have one surgeon and two assistant surgeons. Yet by the summer of 1863, when the big push to form black regiments began, the U.S. Army had already consumed the available medical manpower. Some men were eager for promotion from assistant to full surgeon and were willing to transfer to a black regiment to acquire that rank. Charles Edward Briggs applied for a position with the 54th Massachusetts (the *Glory* regiment). "I think it is desirable that I should have the major strap of Surgeon, before the war is over." He was happy to serve with the black regiment, as "I have been favorable to negro regiments from the beginning, and have no recantation to make if I obtain a position in one."[1] No doubt other physicians took the opportunity of some 138 newly created surgeon positions to garner the promotion in rank and pay. But all too often there were not enough applicants, and the small available pool was comprised of men who had been unable to get a commission elsewhere due to their poor qualifications. The surgeon general set up screening boards specifically for medical appointments to black regiments and claimed that the men had to meet the same requirements as those for white regiments, but there seems little doubt that, in practice, standards were lower.[2]

Historian Paul Steiner has charted the medical staffing of the 65th U.S. Colored Infantry in detail. This regiment never had the expected medical staff of three doctors, rarely had two physicians well for duty at any given time, and for several weeks had none at all. Gathered in St. Louis during the winter of 1863–64, the regiment's first surgeon was mustered in on 9 January 1864. He had graduated from a St. Louis medical college only a few months before; that he entered as a full surgeon with so little experience was a sign of how desperate the colored regiments were to commission anyone with formal medical training. This man developed an incapacitating hernia and left the regiment that summer. A second physician joined on 4 April 1864, and was dead a month later of diarrhea. A third arrived in August, quickly sickened, and left in November. A fourth came a couple of weeks later and lasted until the following August 1865.[3]

The 1st U.S. Colored Infantry Regiment had similar troubles maintaining its quota of physicians. The first appointee refused to come south when the regiment was ordered there. The second found he disliked the duty and neglected it until he was fired. Then a private in the company claimed to have some medical knowledge, so he was appointed surgeon. It subsequently happened that his self-appraisal was not borne out by events, and his principal reason for wanting the position was its easy access to medicinal whiskey. A medical inspector visiting the regiment's camp in September 1863 found the fourth surgeon to be fairly competent but regretted his exhibition of intemperance. On inspection day the man had a black eye and bruised nose, trophies of the previous night's alcoholic revels.[4]

When departmental commanders applied to Washington for more surgeons, all the Bureau of Colored Troops could say was that "there are no candidates available for appointment," and "great difficulty is experienced in obtaining medical officers."[5] The need was particularly acute in the Department of the Gulf during the summer of 1864, as more and more black units were called upon to defend the lower Mississippi River. Commanding general N. P. Banks, frustrated by his inadequate medical staff, first appealed to Washington, but in vain. He noted that hospital stewards had been appointed as surgeons because they were the only medical men available. Hospital stewards were roughly equivalent to pharmacists and were charged with maintaining and dispensing drugs. Banks protested that "well grounded objections were made from every quarter against the inhumanity of subjecting the colored soldiers to medical treatment and surgical operations by such men. It was an objection that could not be disregarded without bringing discredit upon the Army and the Government." Banks pointed out that the acquisition of competent medical officers was not just a matter of benefit to the black troops. "Remonstrances were also made by officers of high rank, commanding white troops, against the appointment of such men, upon the ground that in the exigencies of battle any officer might be subjected to the necessity of surgical treatment by this class of officers."[6]

Banks sent an agent, Dr. J. V. C. Smith, to recruit freshly minted medical graduates from northern medical schools, but this tactic met with little success. Instead, many positions of surgeon and assistant surgeon were filled by medical students just shy of earning the M.D. degree. The qualifications of these men would have been as diverse as the range of medical schools extant in the United States. Some schools required a fairly rigorous course of study, coupled with apprenticeship training, and their near-graduates might have learned quickly and worked adequately to care for the troops. But other schools were little more than diploma mills, and their students would have arrived with little practical knowledge. The troops could recognize the difference. "The doctors visits [sic] them about three times a week, and they do more harm then Good," wrote one black soldier from Texas. "They Poison the Soldiers. they are called doctors but they are not. They are only Students who knows nothing about issueing medicins [sic]."[7] Even tapping the pool of medical students, there were still not enough surgeons to go around, especially competent ones.

The choice of hospital stewards reflected the desperation within regiments to find anyone with medical knowledge to help the sick and wounded men. The hospital steward embodied a combination of pharmacist, hospital administrator, and clerk. He maintained the pharmacy, compounded drugs on the orders of physicians, supervised the hospital staff (except the physicians), ordered supplies, and even at times assisted the doctor in manual tasks such as cupping or applying plasters. There were hospital stewards in each of the general hospitals, and others were assigned to regiments. The stewards came from a variety of backgrounds. Some had worked in pharmacies before the war, where they learned the trade as apprentices. A few had probably studied the subject formally at one of the few established pharmacy schools. Others were medical students, who joined up as stewards rather than miss the excitement of war. Many entered the war with no medical training but enough education to read a little Latin, and keep books. At times physicians or other stewards trained an intelligent private to take over the steward's job or relied on convalescent patients.[8] This would have been easier to do in regiments recruited among northern whites or blacks who had had some opportunities for education. Most of the ex-slave soldiers could barely read, much less decipher Latin medical abbreviations.

Official attempts to prevent the appointment of hospital stewards to surgeon positions met with varied success. General Lorenzo Thomas, the man who was responsible for the recruitment of so many black regiments in the Mississippi Valley, was adamant about this standard. "The subject of obtaining suitable Medical Officers for the Negro Regiments has given me much uneasiness, and I have, in some cases, held back from making such appointments until it became necessary on account of sickness," he told Colonel E. D. Townsend, the assistant adjutant general in December 1863. "The rule laid down by me, and which I have adhered to, was, that

I would not give an appointment to a Hospital Steward, or any person, who was not a graduate of a Medical College. I place the Negro Troops on a perfect equality, in this aspect, with the white troops."[9] Thomas had to repeat this order multiple times in the following months, as he received communications from commanders seeking approval for less qualified men. "No Surgeons or Assistant Surgeons will be appointed in Regiments of African Descent unless it is certified that they are graduates of medicine," he reiterated with exasperation in February 1864.[10]

The very frequency of Thomas's letters rejecting hospital stewards for surgeon appointments reveals that the practice was continuing despite his protests. Indeed, he was probably frequently overruled by the medical boards established to review candidates. In late 1863 Thomas heard from the senior surgeon attending black regiments stationed near Helena, Arkansas, that three of the assistant surgeons in the command were incompetent. Thomas had them ordered to Memphis, to be examined there by the medical board charged with testing candidates for surgeon positions. Thomas found that all three were not M.D.s and revoked their appointments. This action on his part apparently made no difference. "General Prentiss soon thereafter (perhaps not knowing that I had refused to appoint them) assigned them to duty as assistant Surgeons. They were examined, and passed by a Medical Board at Memphis."[11]

Toward the end of 1864 it had become so routine to promote hospital stewards to surgical rank that when a capable steward failed to receive such promotion, the only apparent explanation was the commanding officer's racism. One soldier of the 54th Massachusetts Infantry wrote an indignant letter to the *Christian Recorder*. The regiment had just returned from the battle of Olustee, Florida, to camp at Jacksonville. "One of our captains was sick, and there was no doctor there excepting our hospital steward, who administered medicines and effected a cure." The letter's author described the steward as a "colored man, Dr. [Theodore] Becker, and a competent physician." The regiment's officers, including the recovered captain, got up a petition to support Becker's promotion. But three officers refused to sign, saying they wanted no black officers in the regiment, nor a black doctor. The colonel ceded to their prejudice and did not promote Becker. In this case it seems that the men wanted Becker to have the rank and privileges to go along with the job he was already doing, and apparently doing well. Although the letter writer calls him a doctor, it is unknown whether Becker actually had the M.D. degree or the title was honorary. Still, the indifference to formal educational qualifications was probably typical among officers appointing surgeons to black regiments by the summer of 1864.[12]

African American Physicians for the African American Soldier

There were a few black physicians in North America during the Civil War, and the appointment of these men as surgeons to the black regiments would seem an obvious solution to the physician deficit. In fact, only ten such surgeons served in the Union Army.[13] The story of the most prominent of these, Alexander Augusta, M.D., serves to illustrate the barriers to their appointment within the army. Augusta had been born to a free woman of color in Norfolk, Virginia, in 1825. To satisfy his dream of studying medicine, Augusta eventually emigrated to Toronto, where he graduated with the M.D. degree from the University of Toronto in 1860. In early 1863 Augusta wrote to Abraham Lincoln, requesting an appointment "as surgeon to some of the coloured regiments, or as physician to some of the depots of 'freedmen,'" so that "I can be of use to my race." He appeared before the Army Medical Examining board in March 1863. After some initial discussion of his nationality, Augusta received his appointment as a major and a full surgeon on 4 April 1863.[14]

Elated by his appointment, Augusta fired off a letter to the *Anglo-African* newspaper in New York City, urging other black physicians to come forward. He reported that the board had given him a rigorous but fair examination.[15] Augusta quickly discovered, though, that the prejudices that had driven him to Canada for his education still thrived in the urban North. In Baltimore and Washington, D.C., he ran into trouble on public transportation, where he deliberately asserted his dignity by claiming the seating that his rank deserved. Six weeks after his appointment, on a train in Baltimore, thugs assaulted him, tore off his insignia of rank, and threatened his life. He responded by successfully enrolling the provost marshal's guards in his defense and wrote a lengthy account of the near brawl for the newspapers. He was pleased that "even in *rowdy*

Baltimore colored men have rights that white men are bound to respect."[16]
In February 1864, a driver on a D.C. city railway ignored Augusta's uniform and refused him passage on a designated "all white" train. Augusta took his outrage to Senator Henry Wilson, the radical Republican from Massachusetts who consistently championed the rights of African Americans, especially African American soldiers. Wilson read Augusta's letter on the floor of the Senate and pushed through a bill to make discrimination on D.C. public transit illegal.[17] The *Richmond Enquirer* picked up the story and claimed it revealed the true "mission of the war," namely, the forced acceptance of racial equality.[18]

Augusta fared no better at the hands of his fellow physicians. After an initial appointment as physician to a contraband camp in Washington, D.C., Augusta was assigned to duty as surgeon for the 7th U.S. Colored Infantry, then stationed near Bryantown, Maryland. As one of the first surgeons so appointed, he had seniority over physicians who were commissioned after him. This situation was intolerable to the white physicians who followed in the command. A group of them appealed the injustice all the way to President Lincoln. "When we made applications for positions in the Colored Service, the understanding was universal that *all* Commissioned Officers were to be white men," they wailed in February 1864. "Judge of our Surprise and disappointment, when upon joining our respective regiments we found that the *Senior Surgeon* of the command was a Negro." While the men claimed to be all in favor of the "elevation and improvement of the Colored race," they could not continue in subordination to a colored officer without losing all self-respect. There was no answer to this letter in the archives, but the War Department did detail Augusta to the recruitment center in Baltimore for the remainder of the war, where he examined black recruits.[19]

This solution did not satisfy Joel Morse, an assistant surgeon who felt his way to promotion was blocked by Augusta remaining on the regiment's officer list. In a letter to Senator John Sherman of Ohio, he ranted against the "amalgamation or miscegenation in the appointment of officers," urging Sherman to help him "right this *wrong,* which to my mind is *grave, unjust,* and *humiliating.*"[20] Can it be any wonder that Augusta wrote no further public letters urging black physicians to join the cause? He did continue to fight for the advancement of African Americans in medicine after the war. In 1868 he was one of the founding faculty of Howard University Medical School, becoming professor of anatomy. Two years later he led a group of colleagues in protesting the exclusion from the American Medical Association of black physicians and their associations.[21]

Two other African American physicians were commissioned as officers in black regiments. Little is known about D. O. McCord, who was affiliated with the 9th Louisiana volunteers. John V. Degrasse was eminently qualified, having graduated from Maine Medical School, studied surgery in France for two years, and entered the service with fourteen years experience. He received the rank of assistant surgeon with the 35th U.S. Colored Infantry but was later cashiered for drunkenness on duty.[22] Historian Robert Slawson has found seven additional black physicians who served as contract surgeons or acting assistant surgeons. One, Cortlandt van Rensselaer Creed, was the first African American graduate of Yale University. After receiving his medical degree in 1857 and practicing for a few years in New Haven, he joined the 13th Connecticut Volunteers (Colored Troops). Four of the remaining physicians graduated from known medical schools, while the educational background of the final two is unknown.

There may well have been an unwritten policy, in response to the hackles raised by Augusta's appointment, to put qualified black physicians into positions where they would not offend white surgeons. This was the solution to the Augusta problem, although it was unsatisfactory to some because he retained his regimental position. Other black physicians did not receive commissions at all but were hired to be contract surgeons and appointed to hospitals that served black soldiers or refugees. Contract surgeons threatened no one's rank and could be dismissed without a court-martial.

The army was happy to offer such a position to J. D. Harris in the summer of 1864 but, in spite of his qualifications and evident effectiveness as a physician, failed to commission him as assistant surgeon when he applied in early 1865. Harris does not appear on the contract surgeons list contained in the adjutant general's records in the National Archives, and the work of this black surgeon during the Civil War has not heretofore been recognized by historians.[23] Harris was born in North Carolina, educated in Cleveland, and lived for a while in Haiti. In 1860 he published a book about the latter country, encouraging black Americans to join in the emigration movement to Haiti. The book reveals familiarity with French as well as English literary sources and describes a trip to the Caribbean in the late 1850s.[24] While living in Haiti, Harris took an interest in the local fevers and decided to study medicine. By 1863 he was attending medical school at Western Reserve University in Cleveland and studying under the physician in charge of the marine hospital there. In 1864 he matriculated at the College of Physicians and Surgeons in Keokuk, Iowa, and received the M.D. from that institution. During the summer of 1864, he "asked for the position of Acting Assistant Surgeon," as he was "not willing just then to enter the service for three years, without knowing whether it would be agreeable."[25] Perhaps he, too, knew Dr. Augusta's story.

The army appointed Harris to a position at a hospital in Portsmouth, Virginia, which cared for sick and wounded black soldiers, as well as ill members of the local contraband community. His presence apparently caused quite a stir. "The Union-hating, rebel-loving, good men provoking and gallows deserving Portsmouthians were terribly startled, and with eyes extended, mouth opened, hair on end and hands in pockets, they could be seen in groups, talking very low," reported a correspondent to the *Christian Recorder* in July 1864. When a Union man approached these disturbed folk to inquire as to their distress, he was answered, "Why, says one of the F. F. V.'s 'There's a nigger doctor in Portsmouth, in the capacity of a U.S. Surgeon.'" The correspondent chortled with glee. "It was too much: A negro M.D. upon the sacred soil? They could not stand it. Some of them tried to die, others went in search of the last ditch." In contrast, "Surgeon Harris, with an ability that is second to no surgeon in this department, is rendering invaluable service to the sick and wounded soldiers."[26] Another observer lauded the same hospital, although without mentioning Harris by name. "I must tell you what excellent care the colored soldiers received in the Balfour Hospital, in Portsmouth," wrote one Yankee schoolteacher to her friends in July 1864. "Noble men they are and I rejoice that noble men and women are in charge of them. I have seen in no hospital such *genuine,* direct, and gracious courtesy as the hired nurses in the Balfour show to their colored patients."[27]

Seven months later Harris applied for a regular commission with the U.S. Colored Troops or, barring that, a position working with the freedmen in Savannah. But he remained in Portsmouth until he was examined by a board in Richmond on 4 May 1865. The examination paper Harris wrote survives in the National Archives. His answers to questions about anatomy, materia medica, and diagnosis are straightforward and accurate for the day, but he did not receive the desired appointment.[28] He continued as an acting assistant surgeon but was put in charge of the Howard's Grove Hospital in Richmond. This had been the site of the Confederate smallpox hospital, but after the defeat of Richmond it became a hospital for the colored troops and freedmen. Sanitary Commission agent Ira Russell met him there in the summer of 1865 and found Harris to be "a very intelligent colored gentleman from Cleveland Ohio" whose hospital was "neat, orderly and well located."[29] At some point Harris returned to private life and a practice in Hampton, Virginia. His singular accomplishments must have won him some notoriety, for in 1869 the state's radical Republicans chose him as their candidate for lieutenant governor in a complex political maneuver that split the party. Harris garnered nearly half the votes cast but ultimately lost to his opponent.[30]

Although Harris was a new medical graduate in 1864, and one can question the value of the Iowa degree, acquired with less than half a year of study, he was at least as well educated as the lower tier of white medical graduates, and all available reports of his practice are favorable. Altogether, it appears that the black physicians who served in the war had training similar in quality to their white counterparts and certainly better than the white medical students and stewards being proposed in their stead. If more black physicians of this caliber had served with the African American regiments, the quality of health care might have been better.[31] But it took exceptional bravery and resolution to brook the army's racism and the barriers to practice it created. And, as yet, the number of black physicians in America was far too small to meet the army's needs.

Quality of Health Care

It is difficult to judge the quality of health care that the surgeons appointed to black regiments delivered and the impact of their practice on the death rate from disease. It is likely that many of the white physicians from the North saw their first black patients during the war and may have had trouble understanding their patients and their complaints. As we have seen, medical dogma of the time had taught that the black body was different in its response to illness and therapies, and the white physician new to practice among black men had to find out for himself how far this was true. J. W. Compton, a physician from Owensborough, Kentucky, reflected the common wisdom of his time when he reported that the black man "differs as widely from the white man physiologically and psychologically, as does his skin or hair; hence the importance of understanding his peculiarities." Compton regretted that the surgeons for the black regiments largely came from states without black slaves. "These surgeons are wholly unacquainted with the idiosyncrasies of the negro, a perfect knowledge of which could be acquired only by years of practice among the sick of this race." Compton went on to list a variety of such idiosyncrasies, including response to particular remedies, dietary preferences, and speed of inflammatory illnesses.[32]

Compton's views were not all that extreme for his time, although his racism and assumptions of black inferiority are evident in the particulars. His overall point about physician unfamiliarity with black patients had some relevance. It could be particularly difficult for the novice to recognize rashes and other dermatological findings when unfamiliar with their manifestations in black skin. Eruptive fevers, especially smallpox, measles, and chickenpox, were common in the early months of each black regiment, and the surgeon who could not distinguish them reliably was at a disadvantage. One black soldier disparaged his regimental surgeon, mocking his abilities to diagnose smallpox. "The doctor having charge of the seven cases of small-pox, as is supposed, has acknowledged his ignorance of the true character of the disease, alleging that he never saw black folks with the small-pox, and consequently is unable to decide upon the treatment." Both officers and men were increasingly losing confidence in the surgeon. To further illustrate his unfitness for duty, the soldier continued, "two of the seven cases he pronounced incurable, and to facilitate the matter of interment he ordered two box coffins, and had them placed in the same room with the men, but they stubbornly refused to heed the summons of death and are now doing well." He concluded snidely, "It is hoped that the coffins will find at least one occupant in a Copperhead physician."[33]

When Benjamin Woodward visited hospitals in Arkansas and Missouri in an inspection tour for the USSC, he found many physicians who doubted their ability to diagnose and treat disease in the black soldier. Woodward scoffed at such comments, seeing them as an excuse on the doctors' part for why their patients fared poorly. "Medical officers do not and will not give them proper attention and say 'They cannot diagnose in the Negro.'" He upbraided the physicians for their squeamishness in examining black bodies. "Med officers of the 11th would not and did not examine the men and cases of Pneumonia were undetected because he *would not put his ear to chest of Negro!* Men left sick without care until ready to die, and then *wonder why they died!*" Woodward believed that "if closely watched and promptly treated there would be no more deaths than in White Regts. Negro responds readily to remedial agents, and the same treatment and doses are effectual as with whites." He recognized that the aversion of white doctors was easily communicated to their patients and that this added to the poor therapeutic environment. The "Negro requires kindness and care," Woodward admonished. "They are afraid of Med Officers but as soon as they find they are cared for [they] revive."[34]

Woodward found similarly unimpressive physicians in the hospitals for black troops in Memphis. While examining the reasons for the high mortality rates, he again heard from these officers that the black man lacked vitality. Woodward believed instead that "in the great majority of cases [the high mortality rate] can be traced either to incompetency or carelessness of Medical and Military officers, or to errors in hygiene which might have been obviated."[35] Other observers found similar deficits in their colleagues. One told his wife that many men wearing the insignia of surgeon were "confounded fools," and another remarked that his fellow doctor was a "simpleton." These men would have been incompetent if treating white patients, and their abundance among the black regiments was an indication of how low the criteria were for commissioning physicians in these units.[36]

Even well-trained surgeons could be baffled when trying to bridge the cultural gulf between themselves and the black soldier. In his analysis of camp life in black regiments historian Keith Wilson has described the different cultures that coexisted in camp, that of the white officers and the much larger world of the black privates. He likens this universe as akin to the plantation, where, in spite of the oppressive situation, black slaves created a thriving community with their own music, stories, healers, religious leaders, and so on. In the black regiment there often existed two religions, for example—the Protestant Christian ethic promoted by the official chaplain (who, in some units, was black himself) and a set of subterranean religious practices, often only glimpsed by white authority. There were hints of conjure, voodoo, and magic that would have been anathema to the Christian pastor.[37] In the 1930s many rural southern blacks still understood disease as something that could be the visible sign of another's curse or spell.[38] Similarly, black soldiers may have viewed bodily affliction in this light. They may also have looked for certain herbal remedies or healing rituals when sick, familiar from the slave healers on the plantation and absent in the federal hospital.

The gulf in medical understanding between doctor and patient in comprehending disease may have contributed to inadequate care and poor healing. Charles Rosenberg has argued that mid-nineteenth-century American medicine "worked" because of shared assumptions between physician and patient. Both agreed that when a person became ill it was likely because some essential fluid of the body was out of balance with the rest, or the body had lost its equilibrium within the larger environment. Medical treatments tended to attempt to right that imbalance, removing offending fluids through emetics, laxatives, and bloodletting, or by altering secretions via mercury compounds or drugs that caused sweating.[39] The patient felt out of sorts, the surgeon offered a visible corrective, and both were convinced that bold action had been taken against disease. Where the understanding of doctor and patient was too dissonant, however, this sense of satisfaction and conquest was less likely to follow. White surgeons were impressed at how well men did when tended by the likes of Harriet Tubman and Susie Taylor, black women who carried baskets of herbs and healing authority to black soldiers hospitalized in the Sea Islands of South Carolina.[40] On the other hand, the black soldier who found himself in the hospital being treated by physicians he did not know or trust, who used remedies that were painful, unpleasant, or just unfamiliar, lost the benefit of that shared culture. Is it any wonder that the African American patient reacted with despondency and rejection of his caregivers? "Almost at a rule, when taken ill they turn away from all hope of recovery and die speedily and without appreciable cause," said more than one doctor about black soldiers' behavior under their care.[41]

It was not uncommon for white physicians to characterize the black soldiers as superstitious and gullible. Lacking any way to survey the soldiers themselves about their religious perceptions, we can only glimpse their magical belief system through the racist and scoffing reports of white observers. Benjamin Woodward specifically sought information on this topic when he interviewed physicians caring for black soldiers. He reported that one Arkansas physician found the hospitalized black soldier to be "imbued with fatalism from the Fetish doctrines of his ancestors. He believed that persons and spirits have the power to 'Trick' 'Poison' 'Witch' him, and that when sick he is under the power of some of these fatal influences, that 'his time has come,' and he shall die." Another physician caring for former slaves who were now soldiers likewise noted that "the effects of their 'Fetish' doctrines have a great influence when they are sick. They will tell you they have been 'poisoned' 'Tricked' they often call it by some one." The account of pneumonia was particularly interesting: "In Pneumonia they say they have young scorpions in the lungs which they can feel crawl. When one is taken sick another will draw a little blood and let it stand until the Fibrine separates—If they or the Fibrine thus [illegible word] they say 'That is young scorpions' and medicine can do no good for our time is come and we must die." Both physicians predicted that such patients would not recover unless the doctor could convince them that the conjure was ineffective or had been broken.[42]

Even when deep issues of causation were not involved, it could be difficult for the white doctor to understand his black patient. One Massachusetts physician lamented to his sister that "The ills their flesh is heir to are said to be reduced to three: 'misery in the head, misery in the back, misery in the breast.'"[43] When another Massachusetts physician interviewed physicians at a hospital for black soldiers in Virginia he found universal agreement that it was difficult to comprehend the medical history proffered by the black patient. He made inquiry of all the surgeons at this completely black hospital. "They testify, that at first, before their acquaintance with the negro was very extensive, nearly all his diseases appeared obscure to them," the interviewing physician reported. Two reasons predominated. "*First*—Because of the negro's ignorance of the symptoms and of his obscure and indefinite way of describing them, the terms used often having no relation to the symptoms he wishes to describe," and "*Second;*—Their estimate of the relative importance of symptoms is generally, very erroneous, the intensity of physical suffering being their measure of the danger of the disease. Hence, trivial symptoms are greatly magnified while grave ones are entirely overlooked. The testimony of the Surgeons is unanimous that the negro's opinion of his own case, should be taken cum grano salis."[44] Without accepting the contemporary explanation of this difficulty in communication as due to defects in the black soldier's character, the modern interpreter can be impressed at how often white physicians recorded this gulf between their patients and themselves.

A Very Nice Man, Indeed: The Sadistic Surgeon

Some surgeons went beyond incompetence into sadism and cruelty. Two physicians, irked that some convalescent patients acting as nurses had left the ward without permission, had the men whipped. When they were insolent enough to repeat this behavior, the surgeons had their ankles chained together.[45] Another USCT surgeon named Lyman Allen had to be ordered to treat a man whose foot had been partially torn off by shrapnel. Allen ignored the order because he believed that "a man with a part of his foot cut off by a piece of shell was not suffering much pain." His commanding officer disagreed, and the case went to court-martial, where Allen defended himself on grounds that the patient was not of his regiment and thus not his responsibility. Convicted only of neglect, he continued in the service for another year.[46]

Physicians often drew criticism for suspecting soldiers of feigning illness to avoid duty. "Dr Webbs and Peaz and Saunds are called murderers," wrote one black soldier from Texas. When a sick friend had told the doctor that he was unable to perform his duties as a soldier, the doctor had him "buck and gagged until 12. o.clock and then told the captain to Releace him and Send [him] to the hospital and he gave him Som medicin and the next morning he was a Chorps." Another sick man crawled out of his tent at night "to ease him Self," and "Dr Peaze came up at the time and Cursed him and kick him in to his tent that night he died." A third comrade, suffering volubly in the hospital, drew from Dr. Peaze the sympathetic curse "god damn him if he was a going to die . . . dont [make] So much fuss about it."[47] Dr. Peaze is likely Byron Pease, who served with the 95th U.S. Colored Infantry. One wonders if he is related to Dr. Edmund Pease, who also served with a black regiment in Texas the following summer. Edmund began the war in a Confederate regiment, joined the group of physicians who complained about serving under Alexander Augusta, and was a member of the medical examining board that failed to appoint J. D. Harris as an assistant surgeon.[48]

Other physicians were similarly punitive in their approach to the sick soldier. A private in the 32nd USCT wrote sarcastically of their doctor's attitude. "There is our gentlemanly doctor. He is a very nice man, indeed . . . If they get very sick in their quarters the doctor will order them to be brought to the hospital, where they will not be more than twenty-four hours before they are dead." The soldier believed these deaths were not due solely to ineptitude or the lack of supplies; he accused the doctor of deliberate malignant intent. "Dr. White growls and snaps at the men as if they were dogs, and he says, if the men are not fit for duty, send them to him, and he will soon get them out of the way. And he does put them out of the way, for he says, it is no harm to kill a nigger."[49]

Some physicians apparently felt harassed by the needs of so many sick men. A Sanitary Commission agent witnessed the following scene in 1864. A black soldier approached the assistant surgeon and announced that he was sick. "The answer was 'Clear out with you. You know you have no business here, come in the morning at sick call.'" The Sanitary Commission physician remonstrated with the army surgeon, who replied, "If I were to be bothered every time a nigger thought he was sick I should not have an hour to myself." The agent concluded, "I saw that 'nigger' a few hours after in the corner of a fence with a burning fever. I have asked them time and again, why don't you tell your doctor you are sick? 'Tis no use, he wont believe me if I do.'"[50] Soldiers consequently avoided the unsympathetic doctor, yielding to medical care only when too weak to resist.

What Difference Did a Good Doctor Make?

If the quality and quantity of physicians who took care of the black troops was deficient, what difference did it make? What, after all, could the Civil War doctor do that made a real difference in mortality rates? Quite a lot. He could pull the sick soldier off sentry duty and put him on bed rest. He could agitate for adequate diet, including a supply of vegetables. He could push for the construction of hospitals and provision of quality nursing care. He could complain when the hospital lacked supplies and recognize that the black man needed quinine and opium just like the white. He could push the regimental commanders to maintain well-policed camps with proper sinks that did not contaminate the water supply. The physician who acted as an advocate for the soldiers under his care did not always succeed, but such behavior certainly could benefit the troops. It did make a difference if the surgeon was compassionate and intelligent, as opposed to indifferent, cruel, or absent.

The surgeon worked with the line officers to plan field evacuation, and careful attendance to this duty could make a significant difference in troop mortality. This was doubly true for the African American soldiers, as Confederate troops were known to simply massacre the black prisoners, wounded or not, rather than treat them according to accepted rules of war. One surgeon, Alex P. Heickhold with the 8th U.S. Colored Infantry in Florida, transported wounded black men off the field first for just this reason. A private in the regiment told the *Christian Recorder* that the surgeon "was particular in collecting the colored troops who were wounded, and placed them in his ambulance and pushed for a place of safety." When some protested this triage by race, he said, "I know what will become of the white troops who fall into the enemy's possession, but I am not certain as to the fate of the colored troops." This surgeon was also credited with running a hospital where men were "treated with the utmost attention and kindness" in a house that "once belonged to one of the prominent citizens of Jacksonville."[51]

It was inside the hospital that the surgeon's influence on life and death was most evident. Several factors mattered most immediately in this setting. Was the institution clean? Were dressings changed at appropriate intervals? Did the patients lie in their own excrement? Did they have adequate bedding and clothing to stay warm? Was the food sufficient and of the proper sort to match their condition? Did nurses ensure that the patients received food and drink when they were too feeble to walk to communal tables? A wounded soldier lying in a clean bed, with regular dressing changes, adequate water, a nutritious diet, quinine for his malaria, and opium for his camp diarrhea might well recover. A similar patient lying in dirty straw, consuming only hardtack and coffee, and receiving no medications might well die. Although the surgeon in charge of a hospital did not always directly control the many factors that affected the patient's situation, he could do much to promote a healing environment.

In St. Louis, Ira Russell made an experiment to test this hypothesis. He was impatient with surgeons who claimed that the high mortality in their black wards was due to the inherent weakness of the colored patients. Russell was sure it had much more to do with the dedication and attentiveness of the surgeon. He described one Dr. James M. Martin, who "entered, heart and soul, upon the duties of his ward." In a time of great sickness, Martin carefully tailored diet and medication to his patients' needs. "The result was, that his patients *nearly all recovered,* and the fact was so manifest, that the other Surgeons felt some jealousy and accused Dr. Martin of receiving only the less dangerous cases." So Russell removed Martin from this ward and gave him the worst one in the hospital where mortality was frightful. Its prior surgeon had reported that "it was useless to doctor a sick negro, for he was sure to die, do what you would." "The Doctor [Martin] entered upon his [new] duties with his accustomed zeal and fidelity and, in a short time, the great mortality in this ward ceased and it soon became one of the best wards in the Hospital." Martin succeeded, Russell argued, because he "thoroughly understood the negro characteristics and, in addition to the care and skill he bestowed upon his patients, he knew exactly what psychological influences to bring to bear upon them as remedial agents."[52]

By 1864, when diseased and wounded black soldiers increasingly required hospital beds, there were clear standards for adequate hospitals that were well known among the army's medical hierarchy. In 1859 Florence Nightingale's *Notes on Hospitals* appeared, and its influence on hospital construction and maintenance during the war was pervasive. In common with most Western physicians of her time, Nightingale believed that poisonous air, especially the malodorous mists arising from human waste, rotting wounds, and the fevered body, engendered disease. It followed that the most effective therapy for the sick and wounded was fresh air, and thus she advocated hospital design with an emphasis on ventilation. Most new hospitals constructed during the war incorporated some components of the pavilion plan, in which long slender one-story wards were arranged like petals on a flower, or the crossbars of an "E", so that windows on both sides let in air and sunlight. Ridge vents further promoted ventilation. When older buildings were called into use for hospitals, and such ideal ventilation could not be realized, then physicians sought to purify the air with chemical disinfectants, such as chlorine or acid fumes. Physicians placed heavy emphasis on removing the source of foul odors, hence giving rapid attendance to soiled clothing, changing soaked bandages, and generally keeping hospital inhabitants and accoutrements clean.

This ideal hospital required an attentive administration and adequate manpower to maintain cleanliness standards. Contemporaries referred to "proper policing" or "police regulation" when addressing the hospital managers' adherence to sanitation. Many black hospitals fell short, such as the one in Helena, Arkansas, which "was the dirtiest place inside and the filthiest place out side it was [sic] fallen to my lot to inspect," according to one colonel. "The sick men are dirty their beds filthy and uncomfortable. The police about the hospital, kitchens &c. disgraceful. Not a thing about the whole establishment could I find to mention in favorable terms." Hospitals for black soldiers in Fort Smith and Pine Bluff, Arkansas, and Vicksburg, Mississippi, were similarly nasty, and patients in such institutions were two to three times as likely to die of disease as white soldiers in nearby facilities.[53]

General Lorenzo Thomas found the same or worse conditions in Nashville. In December Confederate commander John Bell Hood had attacked the Federal lines surrounding Nashville, hoping to open a passage to the Ohio River. Instead, soldiers under General George Thomas, white and black, routed and very nearly destroyed the rebel army. The cost was thousands of Union men killed and wounded, and hundreds of those were black. George Thomas was no fan of the black soldier and had resisted using them in combat. Lorenzo Thomas had found the care for black troops inadequate six months earlier; after inspecting the situation he wrote to General George Thomas asking that he order the creation of a black hospital. While admitting that "it is necessary that they should be kept distinct from the white troops," he noted that "the sick are now placed in the same building, a very indifferent one, with the sick laborers and those taken from the contraband camps, and it is difficult for the Medical Officers to make them as comfortable as they should be. The mortality, I learn, is quite heavy."[54] After the battle at Nashville on 15 December, army command finally took steps to remedy the situation. Surgeon-General Barnes ordered Ira Russell to go there, as "the emergency requires the immediate services of every available Medical Officer . . . at Nashville." Russell took charge of Hospital No. 6 and renamed it Wilson Hospital in honor of Massachusetts senator Henry Wilson, a persistent advocate for the black soldier. This hospital for black troops was in fair working condition by January 1865 but not yet completely finished in early March.[55]

One hospital could not accommodate all of the African American casualties from December's clashes, and black troops were also bedded in the much less savory Hospital 16. There Lorenzo Thomas found men in wretched condition. He complained bitterly about "the filthy condition of the wounded, and the bedding. Words of mine cannot describe the utter filthiness of what I saw." Thomas explicitly contrasted the care of two soldiers who had fought side-by-side, one white, one black. The former was cared for well; the latter "suffered in filth for weeks." He witnessed the misery of one "soldier wounded Dec. 15, with leg amputated, [who] was on a bed, the clothing of which had not been changed up to yesterday, and he was still in the dress in which he was carried from the battle-field, everything saturated with blood—and he complained that the lice were eating him up." Thomas asked indignantly, "Had these men been white soldiers, think you this would have been their condition? No! And yet the Black fell side by side of the white with their faces to the Foe." Thomas called for the surgeons to be reprimanded and another man to be put in charge of the hospital. The medical director promised amelioration, but it is not clear that any reform was instituted.[56]

Information about the quality of hospitals for black soldiers is sketchy and anecdotal, making it hard for the historian to draw well-grounded general conclusions. By all evidence hospitals were segregated, either by ward or by treating blacks in separate buildings. Black troops frequently received inferior clothing, tents, food, and military ordnance; it is highly likely that this pattern persisted in the creation and supply of hospitals. The very separateness demanded by contemporary sensibilities meant that existing facilities were off-limits and everywhere new arrangements had to be made. In Memphis, the army chose barracks "without a door, window or floor in any one of them" to house a contraband hospital because "a building would not be taken in the city though the government had scores on its hands because 'a contraband hospital would be a nuisance and the citizens would not like it.'" Black soldiers and black refugees (contrabands) frequently shared hospitals, and the unpopularity of these institutions among the white populace may have been a factor limiting government choices. Again and again, in St. Louis, in New Orleans, in Nashville, black troops suffered in their barracks while government agents dickered over where and whether to establish black hospitals. Choices were then made in a rush, with poor planning contributing to the inherent disdain of many army commissary department agents, who fobbed off shoddy materials on these decidedly second-grade institutions.[57]

It is likely that where hospitalized soldiers had local advocates, their care was better. If there was someone to write letters and generally raise a ruckus over wretched conditions, improvement was much more likely than when no one in the sick man's environment spoke on his behalf. Historian Joseph Glatthaar has accordingly argued that regimental hospitals offered better care for black troops than general or post hospitals. Regimental surgeons were part of the community from which the men came and were likely to be held more accountable by fellow officers and men.[58] The soldier in the general hospital, on the other hand, was cared for away from the community of his messmates and regiment, had no prior bond with the treating physicians, and, given high rates of illiteracy, had limited opportunities to communicate his plight. Where literacy rates among black troops were higher, as in the regiments serving on the eastern coasts of Virginia, North Carolina, and South Carolina, hospital care was probably, on average, better.

Another potential advocate for the sick man was the army nurse. Many Civil War soldiers were nursed by their convalescent peers, who were only slightly less vulnerable than themselves. Some 20,000 women also served in army relief work, north and south, and they offered a different perspective and at times a different voice able to garner attention to hospital conditions. Historian Jane Schultz notes that strong-minded women "sparred with surgeons over the particulars of diet, medical supply, and procedure" in their efforts to improve the lot of patients. "Some even risked their positions by exposing corruption, graft, and neglect."[59] In at least a few instances, white nurses took the part of black men when they witnessed cruel treatment. One nurse reported a doctor for beating a black patient on a trumped-up charge. When the doctor threatened her position, she responded vigorously and observed that he "knew he had gone too far . . . He knew I had power and would use it." Her power lay in social connections that made her words carry weight among the military authorities. Another nurse complained that black men with chronic diarrhea hospitalized in Helena, Arkansas, were being beaten for fouling their beds and alerted attentive ears in Washington.[60]

Schultz has found 420 black nurses listed among the carded service records in the National Archives, composing 6 percent of all army nurses. Another 793 are listed as matrons, a term that could mean something akin to ward or hospital supervisor but also used to describe chambermaids and laundresses.[61] A few black women were literate and left accounts of their experiences as nurses in the war. Susie King Taylor, the best known of these nurses, published her memoirs in 1902. Fourteen in 1862, Taylor followed family members into service with the First South Carolina Volunteers, where she served as a laundress, nurse, and reading teacher for the troops.[62] Sojourner Truth nursed black soldiers in Michigan, and Harriet Tubman cared for them in the Sea Islands.[63] But most African American nurses were illiterate, often drawn from the same populations of free blacks, contraband refugees, and liberated slaves as the black soldiers themselves. While Shultz describes incidences of white nurses praising the bravery and strength of their black male patients, she makes it clear that white nurses looked down on their black nursing colleagues and kept their distance. These African American women were at the very bottom of the hospital hierarchy, desperately poor, and vulnerable to sexual assault. While their diligent service no doubt improved the lot of the individual soldier, they lacked the social power to act as successful advocates.[64]

The African American soldier was frequently ill and was frequently ill served by the army's medical apparatus. Accused of malingering, he was denied access to rest and denied access to the hospital, where he might have received a more nutritious diet. Even when spared confrontation with an inexperienced, indifferent, or sadistic physician, the black patient faced barriers of communication between his perception of illness and its cause and that imposed by his caregivers. Inadequate clothing, food, bedding, medicines, and cleanliness rendered the hospital environment less than therapeutic. The next three chapters turn to specific places and stories, adding depth to this story of malignant neglect.

CHAPTER FIVE

Region, Disease, and the Vulnerable Recruit

Thus far we have considered black soldiers and their overall experiences, without consideration of regional differences. The list of factors influencing black health was largely the same everywhere, but the strength of each influence varied widely depending on place. Black troops were not distributed evenly throughout the Union army. They predominated in areas where troops maintained protective garrisons and were much less common where active fighting was ongoing. No black troops marched with Sherman to the sea, for example, although African Americans did follow his soldiers into Atlanta to secure the city. Black troops were spread up and down the Mississippi River, along the crucial railroad lines in Tennessee, and throughout the coastal areas of North and South Carolina. Accordingly, the black army experience can to some extent be characterized by the places they were stationed.

The diseases intrinsic to a given regiment's mortality pattern were caused by a combination of epidemiological factors, some tied to place and others to the population's immunological status. Thus it makes sense to talk about the disease climates of specific places. This chapter opens with a consideration of the situation in South Carolina and then describes the deadly events in St. Louis. Chapters 6 and 7 take us to Louisiana and the Gulf Coast of Texas, where things got even worse.

Black Troops on the East Coast

The black troops stationed in the coastal regions of the Carolinas were among the healthiest regiments. Given all we know about the importance of advocacy in maintaining health, this is not particularly surprising. The first and second South Carolina infantry regiments, composed of local men who had escaped slavery, were officered by men dedicated to the abolitionist cause and the promotion of black citizenship. Likewise, the 54th Massachusetts boasted officers with a similar commitment to the destruction of slavery, and the large abolition community of Boston closely followed the regiment's progress. The surgeons of these units appear to have been as dedicated to their men as were the other officers.

In St. Louis, Ira Russell would have to beg for access to available hospital beds for ill black recruits. But in South Carolina the surgeon had his hospital prepared before sickness set in. Thomas Higginson remembered, "Our new surgeon has begun his work most efficiently; he and the chaplain have converted an old gin-house into a comfortable hospital, with ten nice beds and straw pallets. He is now, with a hearty professional faith, looking round for somebody to put into it." Higginson worried about the soldiers' lungs, noting that "their catarrh is an unpleasant reality. They feel the dampness very much, and make such a coughing at dress-parade, that I have urged him to administer a dose of cough-mixture, all round just before that pageant."[1] Pulmonary complaints were a major cause of morbidity and mortality among Civil War regiments, black and white, even during the mild Sea Island winters.

The South Carolina regiments do seem to have fulfilled expectations with regard to the killing fevers of summer and fall. The *Weekly Anglo-African* was proud to report in September 1863 that the black troops were doing remarkably well in what was by all accounts a sickly environment. "The immeasurable advantage of sending colored troops to those parts of the South where the heat is so oppressive to white soldiers is emphatically shown by the lists of the deaths in the Port Royal Hospital from July 1 to August 4," the paper crowed. "Of the white troops fifty-seven deaths are reported. Of the black soldiers only nine died during that period. *Every colored volunteer of that number died of wounds received in battle,*" while the white troops died from a variety of diseases, including typhoid and other fevers. The newspaper concluded that this was clear proof that black men were stronger and better able to endure the harsh southern climate than white men.[2]

All was not completely rosy in the East, however. As elsewhere, black troops were put to work on heavy construction projects, building fortifications and preparing campsites, and were worked longer hours than white soldiers. One regimental commander complained to his superior that the fatigue duty assigned his men on Folly Island, South Carolina, was "incessant and trying," so much so that "my sick list has increased from 4 or 5 to nearly 200 in a little over one month."[3] One reason given for the extra fatigue duty foisted on black troops was the very incapacity of white units, in which some 30 percent of the men were down with fever. By November 1863 more black troops were sick than white, in a proportion of 12 percent versus 8 percent, according to one observer. He concluded, "the white troops suffer from fever and dysentery in the summer, the colored from lung diseases in the winter. The Northern colored regts suffer from both."[4] Such a pattern makes sense if the regiments recruited from the region had acquired resistance to local diseases, especially *falciparum* malaria. Although black soldiers from the North might have had some minor advantage in fighting *falciparum* due to inherited hemoglobin variants, the impact of acquired immunity was no doubt greater. Orders restricting fatigue work to reasonable levels helped curb the worst instances of overwork.

It can hardly be said that black troops on the East Coast escaped mortality from illness. At least 5,226 black men died of disease while serving in the Atlantic regions. But compared to the black troops in the Mississippi Valley and Texas, their situation was much better. A wounded man was more than twice as likely to survive a gunshot wound in the eastern theatre, while total mortality from disease was 27 percent of mean troop strength in the West compared to 9 percent along the Atlantic Coast. The accompanying table breaks these figures down further by disease. The data is presented in terms of raw numbers of individual deaths. The mean troop strength in the West was usually about twice that in the East, so if death from disease was distributed randomly between the two locales, then about 66 percent of the deaths should have occurred in the West and

TABLE 5.1
Comparative Mortality among Black Troops by Disease and Region, 1863–1865

Disease Category	Total Deaths	Central Region Total (%)	Atlantic Region Total (%)
Diarrhea	5,238	4,143 (79%)	1,095 (21%)
Malaria	1,481	1,230 (83%)	251 (17%)
Measles	902	804 (89%)	98 (11%)
Pneumonia	4,735	3,723 (79%)	1,012 (21%)
Smallpox	1,535	1,379 (90%)	156 (10%)
Tuberculosis	906	582 (64%)	324 (36%)
Typhoid fever	2,931	2,119 (72%)	812 (28%)

Source: Data in this table was collated from the *MSHW*, 1:651–689. The span of time is from 1 July 1863 to 30 June 1865. The Atlantic region included troops stationed in D.C., Virginia, North Carolina, South Carolina, and Florida. The Central region encompassed Missouri, Arkansas, Tennessee, Kentucky, Mississippi, Louisiana, and Texas. The listed diseases were chosen because as a group they caused 75% of the disease mortality, and no other diagnoses approached them in malignity. Typhoid fever includes typhoid and typho-malarial fevers, as researchers in the late nineteenth-century found that most cases diagnosed as the latter were actually true typhoid. See Vincent Cirillo, *Bullets and Bacilli: The Spanish-American War and Military Medicine* (New Brunswick, N.J.: Rutgers University Press, 2004), on Walter Reed's study of this question. The modern label of malaria is used to summarize five diagnoses: remittent fever, quotidian intermittent fever, tertian intermittent fever, quartan intermittent fever, and congestive intermittent fever. Only the first and last caused significant mortality, and most of these cases were probably caused by *falciparum* malaria. The category of diarrhea includes chronic and acute diarrhea, and chronic and acute dysentery. Tuberculosis is used in place of the older term *consumption*, which generally indicated pulmonary tuberculosis. Pneumonia is used instead of the official category, inflammation of the lungs.

34 percent in the East. Instead, the western rates are all higher, with the single exception of tuberculosis.

These statistics are subject to many caveats. They are death records, so they are likely fairly complete, as opposed to case reports. The army was particular about knowing where its men were, and the paperwork for death in a hospital was hard to avoid. Few diagnoses were confirmed with the sort of laboratory studies the modern diagnostician would expect. Malaria was recognized by fever pattern and time of year, not by microscopic inspection for the parasite, whose role would not be recognized for another three decades. While a few inquiring minds searched for microscopic causes of infectious diseases, in general diagnosis was symptom and examination based.[5] Thus the diagnosis was inflammation of the lungs if there was cough, fever, and crackles on auscultation. If there was chest wall pain, especially on a deep breath, and a rubbing noise could be heard with the stethoscope, then the physician might put down pleurisy instead, even if accompanied by pneumonia. Some cases of typhoid were diagnosed postmortem via examination of the intestinal wall, but most would have been recognized on the basis of fever, rash, diarrhea, and exhaustion. It is hard to know whether some systemic bias in diagnosis was present when comparing eastern troops with western. If eastern physicians were better skilled, did they diagnose more accurately or at least differently? It is certainly possible, but no definitive answer emerges from the records.

The records for the eastern region lump together troops stationed in Virginia, North Carolina, and South Carolina. There were some troops in other postings, such as Florida, the District of Columbia, and Maryland, but most of the men could be found in these three states.

While almost all of the black soldiers along the Mississippi, Ohio, and Tennessee Rivers were former slaves, only about half of the East Coast soldiers had been born in slavery. The other half came from Boston, New York, Philadelphia, and Baltimore, where they would have been subject to urban diseases.[6] Several diseases in the list were dependent on prior exposure and thus reflected the origins of the men. Living in a crowded urban setting, they would have been more likely exposed to measles and smallpox as children or adults. Attentive physicians to the southern troops vaccinated assiduously, although the Sea Island regiments still struggled with smallpox. Typhoid fever was also common in northern cities, where water supplies, especially in urban slums, were contaminated. All three of these diseases provide subsequent immunity, at least for many years after acquisition. And in a setting of continual reexposure, immunity probably remained high. That urban environment also likely explains the tuberculosis rates. Tuberculosis was rife in urban slums in mid-nineteenth-century America. Men who had acquired the microbe as children and young adults may have been mustered into the army with latent disease, only to have it blossom under conditions of malnutrition and excessive fatigue. Rural blacks from the West and South would have been less likely to import latent tuberculosis, although they were equally at risk of acquiring a new case from an afflicted comrade.

The malaria figures raise complex issues about immunity patterns. The African American soldiers recruited from South Carolina and stationed on the southern coast had the advantage of acquired immunity from lifelong exposure to the malaria parasites in the region, as well as innate immunities based on inherited hemoglobin variants. But men who had lived outside of the malarious zones in the years leading up to the war would have been much more susceptible, especially to the deadly *falciparum* malaria. A similar pattern should appear in the West, as men from Missouri, Kentucky, and Tennessee were exposed to *falciparum* in the swamps of Louisiana. So the excess western mortality may reflect just the number of susceptible men exposed. The situation of troops in Louisiana is considered in more detail in chapter 6. It is also possible that the malaria rates reveal a difference in the quality of medical care. Some physicians believed that blacks did not get malaria and that there was no need for quinine to treat them if they did become diseased. An observant and caring physician who eschewed this fallacy might well head off death from *falciparum* through early administration of quinine.

These figures may also represent in part the fact that East Coast recruits were subject to higher scrutiny and more were rejected at recruitment centers than in the Western regiments. The evidence on this point is mixed. Robert Gould Shaw, commander of the 54th Massachusetts Infantry boasted that "we are picking them carefully & shall have a very sound set." He contrasted his lot of healthy men to another of cavalry. "The consequence is that we have an empty hospital, while that of the cavalry opposite, is full."[7] They may also have been a more vigorous group. Comments from physician examiners of the northeastern states, gathered after the war in a volume created by the Provost-Marshal-General's Bureau likewise contain glowing reports about the health of recruits and even the observation that more white applicants were refused for service than black volunteers.[8]

In Maryland, however, source of six East Coast regiments or some five to six thousand men, the standards were much lower. Many of the men were recruited straight off the plantation, with compensation for the owner and freedom for the men as an inducement. It was hard for the recruiting examiner to refuse such men even if they were not physically fit. Once under the aegis of the Union army they wanted to remain there and not return to bondage. The fact that many such were enrolled who were not up to the task is reflected in a policy issued in the summer of 1864 that ordered the recruiting officer in Baltimore, "to cause all such men to be enlisted and mustered into any colored regiment that you may be recruiting at the time; and hereafter such men will not be rejected but will be accepted into the service . . . Muster them on separate 'muster in rolls,' entering upon such rolls the following remark, 'Recruited with a view to transfer to Quartermaster's Department.'"[9] Whether this scheme ever went into effect or not, it reflects the likelihood that men taken directly from slavery may not have been as healthy or strong as those who volunteered in the free states.

All told, the East Coast figures suggest that the black soldiers there received better treatment and, specifically, better medical care. The impressive difference in survival rates from gunshot wounds, for example, allows speculation upon multiple components of mortality. Men with scurvy and other kinds of malnutrition do not heal well, whereas well-fed men are more likely to recover. Surgical skill in tying off vessels, creating flaps, insuring clean bandages, and treating pain all contributed to these figures. Surgeons and other officers who gave proper attention to camp police and the cleanliness of the water supply were more likely to have men who were not wasted by chronic diarrhea after being shot. Much of the suffering in the West could have been prevented. Our story turns now to the streets of St. Louis and one man who fought army bureaucracy to save the black soldier.

St. Louis in the Civil War

St. Louis and Missouri as a whole were hotly contested locales during the Civil War. Slavery remained legal in Missouri until its voters approved a new constitution that abolished slavery in January 1865. At the start of the war, the state's governor favored secession, and the state ultimately sent regiments to both sides in the conflict. St. Louis voted for Lincoln, while the rest of the state supported Unionist Stephen Douglas over states rights Democrat John Breckinridge. But even St. Louis was divided in its sentiments, and only after an armed confrontation between Union men and southern sympathizers, in which the rebels were forced to surrender, was the city secured for the Union. Guerilla warfare erupted sporadically in rural Missouri throughout the war, and when recruiters attempted to enlist slaves, it was not uncommon for their efforts to be met with armed opposition. The proslavery Union man was common in Missouri, and there were few who shared the abolitionist sentiments of New England and the "burned-over" districts.[10]

With its population growing past 160,000 in 1860, St. Louis had the usual civic institutions to care for the sick and poor. People of means received health care in their own homes. The poor might be seen at the St. Louis Hospital, run by the Sisters of Charity, where most of the city's impoverished cholera victims received attention in 1849. There was also a municipal city hospital, which had been established in 1845, destroyed by fire in 1856, and rebuilt just before the war began.[11] The U.S. Marine Hospital cared for sailors of the merchant marine who became ill while docked in St. Louis.[12] It is difficult to calculate how many hospital beds were available in 1861, but there can be no doubt that the number was inadequate for the flood of wounded men who would soon wash up upon the docks of the city. St. Louis was ideally located as a hospital site for western battles. Men could be transported there by water and rail from major battlefields, while the city itself was safely out of harm's way.

In August 1861 some 1,500 wounded men arrived in the city from the battle at Wilson's Creek, Missouri. The available hospital space was overwhelmed, and men lay wherever they could, wearing their bloodied clothing, bullets still lodged in their bodies. Concerned citizens in St. Louis responded by forming the Western Sanitary Commission (WSC), modeled on the U.S. Sanitary Commission, to organize relief for the soldiers. Authorized on 5 September 1861 by Maj. Gen. John C. Fremont, the departmental commander for the West, the WSC raised money, outfitted hospitals and hospital ships, bought supplemental clothing and food for Union soldiers, and later helped care for the black and white refugees who sought respite in St. Louis. The WSC worked hand in hand with the army to expand the city's hospital capacity, in a usually harmonious arrangement. The WSC supplied six hospitals that opened over the next few months, including the post and convalescent hospitals at Benton Barracks in 1862 and a 2,500-bed hospital at Jefferson Barracks. The WSC helped organize a female nursing corps for these hospitals, a move not always welcomed by army physicians. The hospitals experienced waves of casualties after major battles; 1,300 wounded arrived after the battle at Fort Donelson, while the battle of Shiloh sent more than 3,000 to the St. Louis docks. Similar waves followed the battles at Vicksburg and Perryville. In between, the St. Louis hospitals saw a steady stream of men felled by disease and wounds from smaller encounters.[13]

St. Louis also became a major destination for refugees, white and black. As Grant's army took hold of western Kentucky and Tennessee, opened the Mississippi River, and moved into Arkansas and Mississippi, floods of folks fled to the relative safety of St. Louis. "They came on government transports, came by boat-loads, sent by Union generals because they had become a serious impediment to military movements," remembered one St. Louis historian in 1908. "They came also in wagons and carts of wonderful make, and in large numbers on foot."[14] He estimated that some 40,000 white refugees flooded the city during the war, and at times half of them arrived ill, requiring immediate hospital care. A smaller number of blacks made it to St. Louis, but they were equally likely to be sick and, in the harsh winters of 1863 and 1864, frostbitten. The WSC and the U.S. army provided outdoor relief where possible and included wards for refugee care in the vast hospitals created during the war. These destitute people from rural areas added fuel to the fire of epidemics, especially those epidemics that most threatened those previously protected by rural isolation.[15]

Benton Barracks

As the preparations for war accelerated in 1861, St. Louis became a mustering and staging point for armies in the West. To meet the demands for troop housing and training, the U.S. army rented land four miles north of downtown St. Louis and took over the site of the state fairgrounds. There, in August 1861, "workmen began building five barracks, each 750 feet long and 40 feet wide. Kitchen sheds, stables, warehouses and other buildings went up nearby."[16] One soldier recalled that the site was well drained, flat, and quite suited for soldiering. Water was piped in from a reservoir located in an elevated part of the city.[17] The site was used not only as a troop cantonment but also as a parole camp for Confederate soldiers and a refugee camp for whites and blacks. In the first months the area was healthy, but winter brought with it a host of ills.

A physician who arrived with the 3rd Iowa Cavalry in November 1861 depicted Benton Barracks as a panorama of horror and disease. His description began, "The quarters of the rank and file were erected one story high, of rough material, and washed with lime on the exterior. The floors, also rough, were loosely laid and in many instances lower than the surrounding surface." The only ventilation was provided by small apertures cut into the walls "for the admission of air and light, and closed by a door, or a shutter made to slide to and fro." Since "these apertures opened upon the upper bunks, whose occupants at night closed them to exclude the currents of cold air from their persons," the nighttime atmosphere became quite fetid. Adding to the fragrance were emanations from nearby sinks and deposits of kitchen offal, so that "during the whole night, therefore, these quarters were filled with a hot, foul, and poisonous breath, without due means of escape." Into these rough quarters some 30,000 men crammed.[18]

"They were so healthy and brave when they arrived," the physician remembered.[19] But by winter's end, epidemics of smallpox, measles, pneumonia and diarrhea had ravaged the camp. "We had to go through one of the saddest trials of disease we have ever had in our term of service," another Iowan reported.[20] At any given time, a fifth of the men encamped there were sick, and many died, especially those unlucky enough to catch smallpox. Measles spread among them "like the fire upon our western prairies," leaving them weak and prone to other contagious diseases. Some regiments lost forty to sixty men from this disease alone. The army responded by confiscating nearby residences as hospitals, choosing large houses owned by Confederate sympathizers, many of whom had left the city.[21] During February 1862 the assistant surgeon general took over the amphitheatre at the state fairgrounds and converted it into a hospital with 2,500 beds. Different sections of the building served as a post hospital for regiments stationed at Benton Barracks, a general hospital for men sent to St. Louis for care away from their regiments, and a hospital for refugees, including sick women and children. The hospital opened on 1 March 1863.[22]

One week earlier Asst. Surgeon Gen. Joseph Brown had ordered Ira Russell to St. Louis and placed him in charge of the new general hospital at Benton Barracks. After receiving his B.A. at Dartmouth in 1841, Russell graduated from a New York medical college in 1844. By the time the war commenced, Russell was married with three children and practicing medicine in Natick, Massachusetts. He began as commissioned surgeon of the 11th Massachusetts Infantry Volunteers but in 1862 was appointed to the higher level of brigade surgeon. First assigned to direct a hospital in Baltimore in 1862, Russell was then ordered west, where he acquitted himself admirably as medical director of western Arkansas.[23] On 20 February 1863, Russell wrote to his wife, after arriving in St. Louis. He had just received the appointment to the Benton Barracks post and was quite excited. "For some reason I have got the reputation even among the fast men of the west of possessing energy. But I am inclined to think what you call obstinacy passes for energy out here. Next Monday I am to take charge of the Benton Barracks hospital now being built or rather enlarged as it already accommodates 600 patients." He crowed, "My administration in North Western Arkansas was a good thing for me—by it I made a *mark*."[24]

Russell moved quickly to organize his new command. An initial inspection showed that the barracks were filthy, with yards full of litter, privies reeking, the men's quarters and kitchens unclean, and piles of kitchen refuse fouling the air. He ordered a thorough cleaning.[25] By April he had come to realize that a well-organized nursing corps was essential to the efficiency of the hospital. He was fortunate to find as a nursing supervisor Emily Parsons, a thirty-nine-year-old Massachusetts native who had trained for a year in Boston's best hospital before joining the army as a nurse. Russell appointed her supervisor of both male and female nurses, so she reported directly to him. Parsons found him "particular as any general" but also said that he was "good and kind and thorough, and everybody else has to be thorough too." After commenting that Russell was "much beloved here," she admitted "I like his strictness; it is right; but it obliges one to walk very carefully." Parsons had her hands full, what with making daily rounds with Russell, training new nurses, overseeing wound dressing and the delivery of medicines, reading to the men, buying fruits and vegetables with money donated by friends, and generally instilling a dogma of cleanliness and order into her surroundings.[26]

In August 1863 Russell fell victim to "bilious diarrhea" and went home to Massachusetts to recover on an authorized medical leave of absence. When he returned, another man was in charge of the general hospital, and Russell had to make do with the post hospital command.[27] Parsons likewise was felled by sickness and left St. Louis for Cambridge in October 1863. She did not return until March 1864, when she found that "The white are pretty sick, but hardly so much as the colored."[28]

Black Troops at Benton Barracks

Although a small number of black soldiers were treated in the St. Louis hospitals in the summer of 1863, sick troops only began to arrive in large numbers in mid-November, when the First Iowa Volunteers A.D. (African Descent) entered the camp. Sixty men in this regiment were ill and needed immediate medical attention. Russell applied for hospital space to care for them, at a time when some 500 hospital beds were empty. He waited ten days for an answer. "The Negrophobia of the Medical Director had much to do with this unnecessary delay," Russell snorted. This medical director "expressed the very benevolent wish that 'niggers' were all in a certain nameless place." Russell had continual problems getting supplies for these patients, due to the "pettish and factious opposition of the Medical Director," and soldiers went without blankets and other necessary articles.[29] A month later he could still protest, "I am doing all I can for the sanitary condition of the colored recruits and could do much more were I not hampered by Asst. Surgeons U.S.A. who have been allowed to run the medical machinery of this department."[30] The medical director for the Department of Missouri during 1863 and early 1864 was Madison Mills, and, as will be seen, he did not like Russell either. Which specific assistant surgeons plagued Russell cannot be discerned from the records.[31]

The winter months of 1864–65 were deadly for the new volunteers. Historian Paul Steiner has found that some 200 black men, enrolled in the 64th U.S. Colored Infantry, died before their regiment shipped out in March. Circumstances were equally harsh for the 65th Colored Infantry Regiment. It was recruited in Missouri and organized at Benton Barracks. Lt. Col. William Fox, who compiled regimental casualty statistics in the 1890s, recorded that "over 100 men died at the Barracks before the regiment took the field, the men having been enlisted by the Provost-Marshals throughout the State and forwarded to this post during an inclement season." The men arrived "thinly clad, and many of them hatless, shoeless, and without food." As a consequence, many "suffered amputation for frozen feet or hands, and the diseases engendered by this exposure resulted in a terrible and unprecedented mortality."[32]

The situation grew so dire that Ira Russell appealed to Senator Henry Wilson, a friend from back home in Natick, Massachusetts. "I wish to call your attention to some abuses in this Department," Russell wrote Wilson on 20 December 1863. "I am Post Surgeon at Benton Bks the rendezvous of colored recruits. I know from what I see and hear that the state authorities and Provost Marshals do not properly aid in carrying out" orders in preparing for new black recruits. Those recruited straight from the plantation were surrendered by their owners wearing rags; the owners kept the slaves' sturdier garb for their remaining workers. These same owners resisted slave enlistment by punishing the families of slaves who left, putting wives and children to work at the jobs done before by the enlisted men, or selling those families into Confederate states, where they would be out of reach of the soldiers.[33] Russell found that men arrived "thinly clad with ragged and tattered garments, without overcoats and very frequently hatless and shoeless"—and this when the temperatures were dropping routinely below twenty degrees Fahrenheit.[34] "Large numbers had frozen feet, hands, ears, and faces," and nearly all had colds. They arrived at Benton Barracks hungry, sometimes having been kept in railroad cars for two or three days without food.[35]

Senator Wilson, a champion of equal rights for African Americans, was irate. On 12 January 1864 he told the Senate that he had received a letter from an army surgeon in St. Louis. He quoted Russell as saying, "'I wish to call your attention to outrages perpetrated on the negro recruits at this post . . . [who] are harassed and annoyed by conservative copperhead provost marshals and other Government officers before they arrive here.'" Wilson goes on to say that he has "placed the letter this morning in the hands of the Secretary of War." It is evident from the Sanitary Commission papers and Russell's private papers that he was called upon to substantiate his charges in letters to the War Department. No official response has been found, but by March Russell had been made senior medical officer at Benton Barracks, overseeing the post and the general hospitals there. Madison Mills vented his anger in a letter to Surgeon General Barnes, when "on my return from a short leave of absence I find that Sur-

These photographs of Private Hubbard Pryor, 44th U.S. Colored Infantry, illustrate the evolution from slave to citizen via military service. The ragged state of his "before" attire also depicts the inadequate clothing in which most slave recruits arrived at Union Camps. Photographs enclosed in a letter from Col. R. D. Mussey to Maj. C. W. Foster, 10 October 1864, M-750 1864, Letters Received, Ser. 360, Colored Troops Division, RG 94, NARA.

geon G. G. Palmer has been relieved from duty at Benton Barracks" and Russell appointed in his place. Mills charged that Russell had "not the requisite ability to manage a large hospital," but apparently Barnes disagreed. Russell held the position until December 1864, when ordered to Nashville, where he set up a new hospital for black soldiers.[36]

Epidemiology

The deadliest time for black troops at Benton Barracks was the first three months of 1864. This spell of increased mortality reflected the great gathering of men that occurred during that winter, and also the shipping out of most units to other destinations come spring. The 65th U.S. Colored Infantry, for example, boarded the steamer *Sultana* on 12 March 1864 on its way to Louisiana. The 67th U.S.C.I. left seven days later, and the 68th followed within another month.[37] Aside from the deprivations inflicted by the army, the major influences upon the spread of disease were those of the winter season. Hence the diseases characterized by contagion in crowded spaces—pneumonia, smallpox, measles, and meningitis, among others, prospered mightily. Mosquito-borne diseases, on the other hand, were suppressed as their vectors hibernated. Russell may have characterized the barracks' grounds as malarious, but this factor was in abeyance during the snows of January. The various diarrheas and dysenteries persisted but were relatively calmed as the freezing temperatures slowed microbial multiplication in ground water. One disease of crowded, malnourished, wintering people—typhus—was absent among the unfortunate soldiers at Benton Barracks, an absence happily characteristic of the war as a whole.[38]

Pneumonia and Measles

Of the 784 cases of pneumonia Ira Russell counted among the six regiments of black troops at the Benton Barracks Hospital from 1 January to 1 May 1864, 156 were fatal. There were in addition 675 cases of measles, resulting in 130 deaths from pulmonary complications. All told, something like one quarter of the black troops had a bad enough cough to enter the hospital, and one-fifth of those men left the wards in a coffin. It has been known since World War I that a person who has recently had measles is ten times more likely to die of pneumonia, presumably because of a temporary decline in immune function.[39] After noting that the death rate from pneumonia during the first two years of the war among all Union troops was 23 percent, a percentage "much larger than is usual in civil hospitals and private practice," Russell agreed with a proposition put forward by Joseph Woodward that the pneumonia was so fatal because it came as a sequel to measles.

Although many white troops on both sides suffered through measles in the early years of the war, by the winter of 1863–64, they had developed immunity. Many of the new white recruits were immigrants from urban areas, who had already been exposed to measles. It was not so with two groups crowded into Benton Barracks. The prior isolation of the ex-slaves in agricultural areas meant that many encountered the measles virus for the first time in St. Louis. White refugees also filled the hospital. They had fled to the city to escape the guerilla warfare in Missouri and Arkansas or were otherwise displaced by the war. Russell had to care for these people, but he did not like them much. He found that these women, children and elderly men notable "for their indolence, ignorance, filthiness, and inveterate consumption of tobacco. The women were particularly addicted to chewing, smoking, and dipping the pernicious weed. Among this class the disease [pneumonia] was much more intractable and fatal than among the contrabands and colored soldiers." Russell agreed that black soldiers were more liable to contract pneumonia than white soldiers but argued that in comparison to the white refugee, the black man's resistance was superior.[40]

Smallpox

Smallpox was one of the major scourges of Benton Barracks in the winter of 1863–64, when Russell documented some 500 cases among the white and black troops, contrabands, and white refugees.[41] Although preventable by vaccination, smallpox ran rampant due to problems in vaccine supply. Regulations did require that every man inducted into the U.S. army be vaccinated against smallpox upon arrival, unless he could show a scar demonstrating prior vaccination. Few black recruits arrived with the cicatrix on their upper arms. While the smallpox vaccine virus originated from cows, by the mid-nineteenth-century it was propagated from arm to arm, and had acquired a new biological identity. By the twentieth century, this virus would be known as vaccinia, distinct from both smallpox and cowpox.[42] It was the custom in the nineteenth century to manufacture smallpox vaccine by vaccinating children, often poor ones seen in public clinics. When the vaccine pustule formed on the skin, either the milky liquid was extracted and used quickly for the next inoculation, or the scab was lifted and dried. The dried scab could be preserved for some time. When army physicians requested smallpox vaccine, the purveyor sent envelopes of such scabs, which the physician then wetted with water, macerated, and dabbed into newly made vaccination wounds. Given the massive scale of wartime demand for vaccine, the medical purveyors supplied only a few crusts, expecting army physicians to amplify the virus by using vaccinated men as sources for new crusts and lymph.

An alarming outcome emerged from this gargantuan vaccination effort, an unpleasant event known as spurious vaccination. In normal vaccination the patient developed a pustule or even a mild ulcer, about the size of a penny or a nickel. Then the wound healed over and the person was immune to smallpox. This could go wrong in several ways. In some unfortunates the wound grew larger and larger, encompassing the entire upper arm and at times requiring amputation to quell the infection. In others, the inoculation wound took on the appearance of a syphilitic chancre, and the disease itself followed. A third outcome (which could encompass either of the first two misfortunes or stand alone) was that the person would remain susceptible to smallpox.

In the winter of 1863–64, Russell witnessed all three aspects of spurious vaccination. There was widespread panic about smallpox among the men, white and black, and when Russell had difficulty procuring a sufficient supply of vaccine, they began vaccinating each other, with resulting serious ulcers. Russell ordered one regiment of 800 black soldiers vaccinated with crusts supplied by the medical purveyor, and they developed bad arm ulcers and did not acquire protection from smallpox. In one instance a surgeon vaccinated one ward by taking material from a patient with a good vaccine vesicle. "The next day the *same man was taken into another ward* and its occupants were vaccinated. In the same manner he was *taken into all the wards,* it *requiring several days* to complete the work." Each day the surgeon would puncture the man's vesicle (which had closed up over night), and dip points into it to spoon up the milky fluid. Russell reported that "no bad results were experienced by those first vaccinated when the lymph was pure and before the vesicle had become irritated and inflamed, but *some* of those vaccinated on the second day, and the *greater part* of *those vaccinated thereafter,* had local inflammation followed by bad ulcerations with . . . suppuration of the axillary glands . . . [and] pyaemia." Russell concluded that the vaccine matter in the vesicle had become "impure and poisonous by too frequent punctures."[43]

Given the grimy condition of the typical soldier's skin, especially in midwinter, there can be no doubt that the vaccinating surgeons were just as likely to transfer harmful skin microbes, and possibly even syphilis, as they tried to convey the vaccine virus. The possibility that army surgeons had conveyed syphilis to enlisted men aroused so much controversy that a local committee, chaired by a Dr. Hammer of St. Louis, investigated the situation. Committee members examined more than 200 soldiers. They concluded that "the condition of the patients was produced principally by two causes, irregular and improper vaccination and exercise. Many of the soldiers vaccinated themselves, using for the purpose rusty pins, irregular incisions, etc." This initial mistake was exacerbated by excessive drill and other exercise, and when the men were properly rested and treated, the cases of undue ulceration resolved. The committee concluded that the vaccinations had not, after all, transferred syphilis.[44]

Unfortunately, many of these vaccinations also did not prevent smallpox. In the series of procedures described above, where one host was used repetitively, only those men in the first ward treated were protected from the disease. Much of the officially supplied vaccine matter was apparently also inert.[45] As a result, many of the men in Benton Barracks contracted and died of smallpox. Others who had been vaccinated in St. Louis helped fuel epidemics elsewhere after their regiments had moved to different posts. Dr. Joseph Smith, medical director for the Department of Arkansas, noted in his report for 1864, "The vaccine virus furnished to the Army was very unsatisfactory in its results. Very many vaccinations were unsuccessful and some produced spurious sores followed by constitutional effects." He specifically mentioned that "the first cases of spurious vaccination coming under my notice, occurred in the persons of recruits vaccinated at Benton Barracks, Mo., and arriving in this Dept. in May."[46]

Smallpox was rare in the southern states. While its occurrence was sporadic but persistent among northern troops throughout the war, the disease did not appear among the Confederate soldiers until returning prisoners of war brought it in the fall of 1862.[47] Smallpox was a disease kept alive in urban areas, and the southern soldier and the southern slave instead lived mostly an agricultural and isolated existence, with rare exposure to smallpox before the war. Smallpox is directly and rapidly contagious, and the close quarters of the barracks during those frigid winter months made for an ideal site of propagation. Forty-one men of the 65th U.S. Colored Infantry regiment died of smallpox, and probably at least twice that number survived the disease.[48] Although the official records provide information on white troops for the Department of the Missouri, black troop data was listed only for the "Central Region," an area that included Missouri, Tennessee, Arkansas, Louisiana, and Kentucky. Morbidity and mortality from smallpox among all black troops peaked in the spring of 1864 when there were 8 cases per 1,000 troop strength, with 30 percent mortality. Overall 6,713 cases of smallpox were reported for black troops; 2,303 of these occurred in the central region during the spring of 1864.[49]

Meningitis

It was Christmas Eve 1863, and John Tyler had been feeling somewhat punk for a few days. Still, he ate breakfast and drilled with his fellows in the 1st Iowa Infantry [of] African Descent. Lunch had no appeal, and instead he lay on his bunk, suddenly quite cold. His comrades found him there, with his neck and back muscles in spasm, delirious and completely helpless. Carried to the Benton Barracks hospital by friends, Tyler was found to be almost pulseless, and Acting Asst. Surgeon J. M. Martine did not expect him to survive the night. Treated with quinine, capsicum, alcohol, and hot bricks to the feet, he slowly improved, so that by 25 March he had recovered his mental faculties and had only residual neck and back pain. Private Tyler was one of the first cases of cerebrospinal meningitis among the black troops at Benton Barracks. Forty-nine of his comrades would likewise fall ill, but only half of those would live to convalescence. Many of those who died underwent autopsy at the hands of Drs. Russell and Dwelle, who found pus and vascular congestion upon dissecting the brain and spinal cord.[50]

On 11 April, William Allen of the 12th Missouri Cavalry drank six bot-

tles of ginger-pop and feasted on a variety of treats from the camp sutler.
By evening he complained of pains in his head, back, neck and throat, and
went to the hospital. He rapidly progressed to delirium and kept ex-
claiming "Orderly, let me alone, I can't drill today." It took a nurse in constant attendance to keep him in bed. Over the next few days, he remained
incoherent, with muscle contractions so violent that only chloroform
would give him rest. On the ninth day he died. At autopsy physicians
found his spinal cord, brainstem, and cerebellum coated with pus, and
the meninges highly inflamed. Russell only mentioned four white patients with meningitis and did not give an overall number of cases or
deaths.[51]

Official medical records of the war are not very helpful for sorting out
how frequent and severe meningitis was during the conflict. The diagnosis itself does not appear; instead, there are categories for inflammation
of the brain and its membranes, and of the spinal cord. Cases seen in clusters, especially ones that were autopsied, were probably more accurately
recognized and reported than isolated incidences of the disease. Diagnosis based on microscopy would not be possible for another three decades.
Further, the numbers that do exist are suspect in themselves. For white
troops there were more deaths from inflammation of the brain and its
membranes during the year ending 30 June 1864 than there were cases
of these diagnoses. It is thus likely that cases of meningitis were sub-
stantially underreported. Still, it is worth noting that if this cluster of brain
and spinal cord inflammation is summed up and used as a proxy for
meningitis, the white troops had .001 deaths from meningitis per 1,000
mean troop strength for this year, while black troops had .004 meningitis deaths, or four times as many.[52]

The disease had been very uncommon in the first years of the war, but
by 1864 it had popped up in a number of areas, including among the civilian populations of Massachusetts, Pennsylvania, Indiana, and Illinois.[53]
Epidemics broke out at New Bern, North Carolina, and in the penitentiary
at Little Rock.[54] The increased crowding and decreased resistance of black
troops, in evidence at Benton Barracks, may have made them more vulnerable to this virulent infection. Other black men, impressed for work
on Confederate defenses or in crowded labor camps working on federal
fortifications, also suffered high epidemic mortality during 1864 and into

the winter of 1865, although it is difficult to say with certainty that these outbreaks were caused by meningitis. The spotted rash was harder to discern on dark skin, but this epidemic was marked by fever, rapid onset, high mortality, and headache.[55]

Reaping the Whirlwind

Amid this carnage Russell struggled to keep the hospital in order, care for the many sick, and find answers to the reasons for the widespread illness. In the early months of 1864, he performed at least 800 autopsies of black soldiers. He was proud of this accomplishment and mentioned it frequently in reports and papers. Historian Michael Sappol has described the importance of anatomy in the designation of the educated physician in the mid-nineteenth century. Knowledge of anatomy distinguished the learned man from the ignorant; the examinations of physicians for Union army appointments consequently emphasized anatomical detail. In a war where surgical amputation was a major function of physicians, a detailed grasp of the arrangement of arm and leg blood vessels, nerves, muscles, and bones was essential. Russell was pursuing scientific answers to questions about black men and disease, using the foremost research tool of his day. In applying in the summer of 1865 for a job with the Sanitary Commission, which would send him to explore questions about black soldiers' health and disease, Russell touted his autopsies as evidence of his qualifications for the task. He was quite proud of his work.[56]

There is a seeming paradox between Russell's earnest efforts to promote the welfare of the black soldier and his eager willingness to desecrate their corpses after death. There is no indication that he was aware of any dissonance, although the only mention of the autopsies comes in public writing. If asked, he might well have answered that he interrogated the dead to find information to help the living. But it is no doubt also true that his medical education and experience had evolved within a professional sphere that took the dissection of the poor and ethnically other (black, Irish) as a given. "In Baltimore the bodies of coloured people exclusively are taken for dissection, 'because the whites do not like it, and the coloured people cannot resist,'" commented Harriet Martineau during her tour of the United States in the 1830s.[57] Even at northern medical colleges, most of the antebellum cadavers dissected were black, and some had even been shipped from the South for the purpose.[58]

Russell trained at the Medical Department of the University of the City of New York, the ancestor to today's New York University Medical School. Sappol lists this institution as one of two in New York City which boasted of abundant "anatomical material" in recruiting catalogues during the years that Russell was a student. Many of these bodies were black. The availability of such cadavers for education was the mark of a superior medical school, and autopsy experience marked the superior student and practitioner.[59] Whether rumors about the autopsies in St. Louis reached the men remaining in the wards is unknown. These men were certainly vulnerable, given that their families were usually far away, out of contact, and many of them still in slavery. Most of those families lacked the literacy or the knowledge to protest the autopsies or to claim the bodies for burial.

What questions did Russell seek to answer in plunging his scalpel through all that black skin? He organized his autopsy information in tables, which granted each body one long row. The information included height, weight, degree of whiteness, and age. Then the various organs were described, including the brain, heart, lungs, stomach, liver, spleen, and intestines. There was a column on the far right for remarks. Documents recording 489 autopsies survive in the U.S. Sanitary Commission papers. The fate of the rest of the charts is unknown, and if Russell ever wrote a summary description of the autopsy findings, it has not survived. Some results appear in Sanford Hunt's 1869 essay on the black soldier, which quotes Russell at length and uses material now in the Sanitary Commission papers. Hunt's handwriting appears on the cover sheet of the autopsy graphs, with a line saying that the brain weights had been tabulated.[60]

Perhaps the most peculiar column in the grid was labeled Color. Although the overall series is titled "A Series of Colored Soldiers," 5 percent of the bodies are labeled white, and presumably came from the white refugee population that was cared for at Benton Barracks. There are two subsets of graphs, with the first set of bodies numbered 1 to 189, and the second from 1 to 300. In the first set, all of the men are labeled by percent whiteness, with fractions such as 3/4, 3/16, or 1/8, followed by ditto marks that lie under the word white. There is no information about how

Russell decided a man was, say, 3/16 white. Almost none of the soldiers listed in the first set were zero percent white. In the second roster, recorded with a different handwriting and slight alteration in the columns, there are many men listed as black, as well as others with the percentages of whiteness. The recording of mixed race by degree reflects the high interest in the health of such men, but nowhere have I found Russell or others drawing conclusions from this set of numbers.

In the modern hospital, physicians seek autopsies of deceased patients in order to confirm diagnoses and learn from medical mistakes. Only once in these records is it apparent that Russell was surprised by his findings, in a case where he recorded that a man had *"tuberculosis, not pneumonia."* Although prewar discussions of black men as frequently scrofulous predominated in the medical literature, Russell found very few signs of active or prior tuberculosis in his early records. As time went on, however, more and more cases came to the table, a likely sign of spreading disease amid the crowded conditions of Benton Barracks.

Much of the data Russell collected seems pointless, and it is not at all clear why he chose to preserve it. He weighed the brains, and Hunt would later publicize average weights calculated from this data. He found smaller brains among blacks than whites, interpreted as a sign of mental inferiority. But to what purpose did Russell weigh the lungs, heart, liver, and spleen, or reckon the volumes of the arachnoid cavity and the ventricles of the brain? The answer appears to lie in directionless empiricism, the gathering of data because it is there, not because it addresses any particular hypothesis or question. That happened often during and after the war, as physicians struggled to make sense of mountains of experience.

In some ways, the situations of black soldiers in South Carolina and St. Louis represented the best circumstances for the sick that occurred during the war. Their health care was in the hands of caring, competent physicians who were dedicated to helping them survive wounds and the rigors of camp life. But even those dedicated officers could not overcome the pervasive racism of the northern army, manifested by a bureaucracy committed to treating the black man as disposable and the actions of certain physicians who lacked compassion for the men. And they could not begin to overcome the inevitable outcome of bringing large numbers of rural black men together into crowded camps where infections spread with heightened virulence amid virgin soil. The troops leaving Benton Barracks for Louisiana in 1864 and the East Coast troops shipped south to Texas in the spring of 1865 found out that circumstances could be even worse than they had been in the diseased camps they left behind. In these desolate outposts, compassionate advocates were mostly lacking, and mortality rates soared.

CHAPTER SIX

Louisiana

"The Fourth of July 1863 was the most memorable Independence Day in American history since that first one four score and seven years earlier," historian James McPherson has written.[1] It was indeed a momentous day. Twenty-four hours earlier Robert E. Lee had abandoned his plans for a northern invasion and rapidly retreated from the deadly fields of Gettysburg, Pennsylvania. And on the glorious fourth itself, Ulysses S. Grant accepted the surrender of Vicksburg, the last major Confederate fort blocking the Mississippi River. Five days later at Port Hudson in Louisiana, the besieged Confederates, subsisting by then on rats and mules, surrendered to Union general Nathanael Banks. "The Father of Waters again goes unvexed to the sea," announced Abraham Lincoln. The Confederacy had been divided in two, leaving Texas, Arkansas, and Louisiana cut off from the other states in rebellion.[2]

The Union now had control of the country's major river but not necessarily the land on either side of its banks. Forts lately in Confederate hands had to be manned now by Union troops and urban centers on the river garrisoned by even more Union troops. The region captured was one of the most insalubrious in the United States. As the Mississippi River winds through the South, it enters a delta region marked by frequent flooding, swamps and bayous, and levees that often give way. This large amount of free water, accompanied by the South's torrid summer temperatures, created a land perfect for breeding the mosquitoes that transmitted malaria. In New Orleans, the high water table meant that wells were impracticable, so urban residents built cisterns to store rainwater for drinking. The yellow fever–carrying mosquito found these receptacles ideal for their reproduction. These two diseases contributed mightily to the region's reputation for ill health and early death.[3]

Physicians knew well that these fevers were particularly prone to strike strangers and that local inhabitants were protected by their years of seasoning. So it was with dread that Union commanders considered stationing men along the lower Mississippi within this hostile environment. Six months after Union gunboats had secured the lower Mississippi from Natchez to the Gulf, Colonel S. B. Holabird lamented the ill health of white troops in the region and recommended the experiment with black soldiers be tried. Writing to Major General Banks, then commander of the Department of the Gulf, he advised, "The Forts below the city, as wood & Pike on the lakes, St Phillip and Jackson on the river, with Ship Island & land Spits now held by white troops, it appears to me, might be two thirds, or more, garrisoned by negroes, with both propriety and policy." His reasons were straightforward. "The whites suffer terribly by disease and the men become weak. The negroes do not so suffer, are very strong and would fight should it ever be necessary." While he recognized that some white troops would be necessary, he thought their danger could be minimized by frequent rotation in and out of the region. "I think it will save our men and make the colored corps useful," Holabird concluded.[4]

Banks saw the wisdom in this plan, and Secretary of War Edwin Stanton did as well. In January 1863 Stanton ordered General Daniel Ullmann to raise a brigade of black soldiers in Louisiana and authorized the creation of the Corps d'Afrique in the summer of 1863. General Henry Halleck told U. S. Grant in late March that the policy suggested by Holabird in December was now the established plan of the Lincoln administration. "It is the policy of the government to use the negroes of the South so far as practicable as a military force for the defence of forts, depts, &c," Halleck wrote. "If the experience of Genl Banks near New Orleans should be satisfactory, a much larger force will be organized during the coming summer; & if they can be used to hold points on the Mississippi during the sickly season, it will afford much relief to our armies."[5] Although the employment of black men on (now Union government controlled) plantations was a competing need that complicated federal policy, by the fall many regiments of black soldiers were manning the defenses of New Orleans and surrounding areas. Their experience would show that, contrary to Holobird's prediction, the black soldiers could, indeed, suffer in the Louisiana swamp country.

The Regiments in Louisiana

The first black regiments to be stationed in Louisiana were those of the Corps d'Afrique, a group that originated with the Louisiana Native Guards. Prominent African American leaders in New Orleans organized the first of these regiments from the city's large free black population in the first year of the war. Two regiments of these Native Guards fought at the battle of Port Hudson, Louisiana, in late May 1863, earning glory and honor for themselves and their race. In June these regiments were folded into the new Corps d'Afrique being organized under General Daniel Ullmann. Ultimately composed of some three dozen regiments, the Corps d'Afrique served mostly in Louisiana, although a few regiments were dispatched to other sites on the Gulf Coast as needed.

In the spring of 1864 the Corps d'Afrique units were renamed as regiments of the U.S. Colored Infantry, in a reorganization scheme that brought all black regiments into one national system of numbering and control. That spring another nine black regiments arrived in Louisiana. Some came from as far away as Philadelphia and New York, but most came from the nearby states of Alabama, Missouri, and Mississippi. These regiments rarely held the traditional complement of 1,000 men; 500 was more common, and initially the Corps d'Afrique regiments were designed to have only 500 men. So, all told, something like 16,000 black men served in the swampy lands of southern Louisiana in 1864 and 1865.[6]

There were white soldiers in the Department of the Gulf as well, and in fact their numbers grew after the impulse to recruit black troops for the region began. The mean troop strength for the department, which included Louisiana and the Gulf Coast to the west of Pensacola, averaged 26,199 white men for the year ending 31 June 1863, but grew to 47,035 for 1863–64, and dropped slightly to 45,629 for the year ending 31 June 1865. The official mortality figures for causes other than battle casualties may show the influence of the arrival of the black troops, however. The

Black troops on watch in Louisiana. *Frank Leslie's Illustrated Newspaper,* 7 March 1863.

death rate for noncombat casualties in the first year was 7.4 percent. This rate dropped almost in half over the next two years, to 4 percent in 1863–64 and 3.6 percent in 1864–65. Typhoid and diarrhea/dysentery diseases remained major killers. The malarial diseases did drop slightly in the causes of mortality, moving from 0.6 percent of white soldiers' deaths the first year to 0.4 percent in the final one. This drop may reflect the growing acclimation of the white troops, or it may in part reflect their decreased environmental exposure as black troops took over posts previously held by whites.[7]

Union forces in southern Louisiana overran many plantations, and large contraband camps quickly formed near and in New Orleans. Fearing that large numbers of idle blacks threatened the public order, Union commanders wanted to either lure the blacks back to work on the plantations or induct the men into the army. Conditions on the army-supervised plantations did not differ all that much from slavery, and some of the black men were forcibly conscripted into the army. The Confederate press seized with glee on stories about contraband misery. Thomas Moore, Confederate governor of Louisiana, told the *Richmond Examiner* that the "freed" blacks in his state were living under conditions of total destitution and tyranny. "The men are driven off like so many cattle to a Yankee camp, and are enlisted in the Yankee army. The women and children are likewise driven off in droves, and put upon what are called 'Government Plantations'—that is plantations from which the lawful owners have been forced to fly, and which the Yankees in Louisiana are cultivating."[8] The propaganda slant of the story was obvious, but other sources indicate that it had some basis in fact.

In late 1863 the president of the Western Sanitary Commission, James E. Yeatman, issued an appeal for funds to help the freedmen throughout the Mississippi Valley. He composed a pamphlet that reported on their condition with sympathy, although one might suspect some bias on his part in making their condition appear as needy and worthy as possible. Still, he had the following to say about conditions at Young's Point, Louisiana, a site near Vicksburg. The camp housed about 2,100 people, and was under the command of D. L. Jones of the 9th Louisiana Corps D'Afrique. "There appears to be more squalid misery and destitution here than in any place I have visited. The sickness and deaths were most frightful. During the summer from thirty to fifty died in a day, and some days as many as seventy-five."[9] Others put the mortality rate in contraband camps in the Mississippi Valley at 25 percent.[10] It was from this population, which had already endured months of hardship and disease, that recruiters drew the majority of the men who would make up the Corps d'Afrique. Many were too disabled to serve well. One army inspector claimed that an entire brigade of black troops in Louisiana was composed of a "very inferior personnel of . . . men" who had been "recruited without proper medical examination."[11]

Securing and Guarding Port Hudson

In late May the famous battle of Port Hudson demonstrated the valor of the Louisiana Native Guard that fought along side white regiments. That battle did not succeed in reducing the rebel fortress, however, and a siege ensued. Confederates killed 500 men in battle and wounded another 2,500. Not until 9 July, when the rebels holding the fort heard that Vicksburg had finally fallen to General Grant, did the fort capitulate. Union troops occupied the fort and set to work making it stronger and even more impregnable. They also were finally able to bury the dead from the battle six weeks earlier, for the rebels had refused to allow Union burial parties safe passage to inter those who had fallen.[12]

Much of this work was done by members of the Corps d'Afrique. Their officers kept them at hard labor for longer hours than any would dare assign to white men. This excess labor took its toll. Senator Henry Wilson pled their case on the floor of the Senate in June 1864. "From Cairo to New Orleans, the Mississippi at nearly every point is guarded by colored soldiers . . . These soldiers have been forced to work night and day to throw up fortifications and do the drudgery of the Army. They have suffered, toiled, labored as no troops in the service have toiled and labored. We have lost unquestionably a great many of them on the Mississippi, as we have on the Atlantic coast, for wherever there is a colored regiment, and there is drudgery to do, the drudgery is put upon them."[13] One U.S. Sanitary Commission inspector noted that the "immense fortifications at Port Hudson" had been begun "early in Sept 1863—at a time of the year when Malarious Fevers are very prevalent," and high rates of disease had predictably followed.[14]

One inspector who visited these men in the fall of 1863 found them poorly housed and fed. "Tents are torn and old and offer little protection from rain," he wrote. Another units' tents were so worn out that he suspected they had been turned in to the quartermaster by a white regiment. The men were dressed in rags and had holes in their shoes. They lacked fresh meat and vegetables. Most of the men had been contrabands, and he found many men unfit for duty, having been mustered in without exams. There was great dissatisfaction among them about a promised bounty that failed to materialize and their inadequate pay, which left their families wanting. One regiment had enrolled 600 men since July, but of that number 100 were dead by October "from measles and miasmatic fever."[15] The regiments had to endure the same seasoning as the men in Benton Barracks when it came to measles. Although the men from New Orleans had probably seen it as children, many of the soldiers from rural plantations would have met the disease for the first time in the Corps d'Afrique. They suffered the usual depredations from the disease, including the susceptibility to pneumonia.

1864

By the spring of 1864 the Corps d'Afrique regiments had been joined by others, some coming from Benton Barracks in Missouri. These men who had already endured weeks of infectious diseases were now introduced to the fevers of southern Louisiana. One colonel in the 67th U.S.C.I. wrote a friend that "the sick are poorly provided for." The overall tone of his letter is sympathetic to his black troops, and he is particularly angered by news about the massacred black men at Fort Pillow in Tennessee. As to his men, they know what to expect under Rebel capture, so their motto is "death before surrender." Unfortunately for his men, however, a stronger foe was at the gate. "The 67th [U.S.C.I] is quite unhealthy. From one to three men die daily," reported the colonel. "The negroes so far cannot stand the South any better than white men."[16]

The most detailed account of life in a black regiment stationed in Louisiana in 1864 and 1865 comes from the letters of Duren Kelley, a white Minnesota man who rose to the rank of lieutenant in the U.S. Colored Infantry.[17] In some ways, Kelley has a winning personality. He writes sweet endearments to his wife and sentimentally describes a kitten who ruffles his papers and rides around on his shoulder in camp. Yet one gets the impression that he is much like those officers described by a disgusted officer in the Corps d'Afrique. "Men who have been assigned to this corps to hold responsible positions as officers, are totally unfit. They do not believe in the colored man, and nothing troubles them so much as to be considered an *Abolitionist* . . . They are here . . . because they can thereby wear shoulder straps, and have their pockets liberally supplied with greenbacks."[18] Kelley applied for an officer position in the black units precisely to achieve promotion and better pay; he had no sympathy for the black man. At one point he says he prefers Copperheads and proslavery men to the men under his charge, and elsewhere he gleefully reports that he is a disciplinarian who has no qualms about hanging a delinquent soldier by his thumbs.[19]

Given Kelley's disinterest in his men, it is probably not surprising that he reports on the high mortality in his regiment with total detachment. Nowhere does he suggest that any intervention, by himself or others, could improve the situation. He laments the absence of vegetables from his own plate but never seems to consider writing the Sanitary Commission for vegetable supplies. He mentions that he lives in a house built of wood and that the men are under canvas, with no concern about that disparity. Kelley is convinced that the key influence is environment, and that for the men to undergo seasoning is the only way to improve their health. He sounds like Mark Tapley in Charles Dickens's *Martin Chuzzlewit*, who says that that "we must all be seasoned, one way or the other. That's religion, that is, you know."[20] In the spring of 1865, when 107 new recruits were assigned to his regiment, Kelley said resignedly, "This will make a very good regiment again but they—the recruits—will half of them die off this summer. The regiment is very healthy now and those who have been here one summer will stand it. I dread to see the hot weather coming again."[21] There is a certain fatalism here, the assumption that nothing could prevent the automatic winnowing by the toxic environment.

Kelley's men arrived in Louisiana in late March 1864. They camped in the battlefield before the Port Hudson fort, which still bore ample signs of the previous summer's engagement. "I wish you could see the battlefield between here and the plantation—Trees, rivers, with shot and shell—Broken gun carriages, muskets, cannon shot—pieces of shell, and in fact all the implements of war strews [sic] in every direction," Kelley wrote his wife. But other sights were even more harrowing. "And above all the graves of the fallen of both armies hundreds upon hundreds, all to prove how desperate must have been the struggle, and here we are holding peaceful possession of the dearly fought prize."[22]

In May the health of the regiment had apparently improved, and so had the diet, as spring produce came to market. Kelley reported that "we are living very well. Had for dinner today greens, lettuce, potatoes, dried beef, beans, cornbread. Such kind of grub is good enough for soldiers, isn't it."[23] He does not say what the men were eating but does comment on the abundance of their food. Kelley told his wife, "I think that the general health of our regt is better than it has been. Though there has been considerable sickness still among the niggers caused principally by eating too much. They are inveterate hogs and will frequently eat enough to kill themselves. For such fellows I have no sympathy."[24] Given Kelley's attitude, it is impossible to know what this illness caused by supposed overindulgence might have been, but perhaps there was indeed some respite from the ill health that had dogged the regiment since its creation in St. Louis.

But they were camped on a graveyard, and in the June heat the site became unbearable. On a day when Kelley noted that the men had finally been issued new rifles, he also let his wife know that they would soon be moving camp to a place by a creek in the woods. "We are glad to get out of the sand and the everlasting stench arrising [sic] from the graves of dead men, half buried in the siege. It is intolerable nights and we are afraid of the cholera if we stay in our present camp." One wonders how his wife responded to the next sentences. "I am not sure, but there is a grave under my bed. I have noticed a mound under me that hurts my back nights and have now moved my bed. Port Hudson is one vast burying ground with numerous bones of dead men unburied still."[25] Kelley's men sank their wells through that graveyard, and it had the high water table characteristic of southern Louisiana. It is no surprise that typhoid fever and dysenteries did well in such an environment.

Most of the black regiments camped at Port Hudson moved ten miles upriver to Morganza in the last days of June. In the 65th U.S.C.I. regiment only 526 men were fit to march; 139 stayed back in the Fort Hudson hospital.[26] Kelley was pleased by the move, which he felt put his regiment in healthier surroundings. "The white troops are leaving this place as fast as they can get boats to take them and the niggers will soon be left in all their glory," he recorded. "We are building a fort on the bank of the river and will soon be moving inside the works. There is only one bridge [brigade?] of colored troops here besides our own though there will probably be some more out here or some of the white troops left. I think this place is preferable to Port Hudson. We don't lose as many men here as the former place."[27]

Kelley's pleasure in being at Morganza on the fourth of July soon was ruined by his own battle with the local fevers. By 7 July he was back in Port Hudson, in the hospital. He made fairly light of it in a letter to his wife. "Since I last wrote you I have had the fever which I succeeded in breaking up in two or three days and now am as good as new," he wrote her.[28] But in January 1865 he remembered that time as more dire, saying of a departing friend, "He and I occupied the same room in the hospital at Port Hudson last summer and I never shall forget an expression of his one afternoon as we were laying there on our beds nearly dead, hardly able to raise our heads."[29] He spent much of the summer sickly and was pleased when cooler weather allowed him to gain strength and put some pounds on his wasted frame.

The black troops were working excessively in the tropical heat. While Kelley mentioned airily that "we are drilling about three hours a day and the rest of the time take the thing as cool as possible in this latitude," others described a much heavier workload.[30] Lt. John L. Rice reported that "my whole available command [the 75th U.S.C.I.] has been kept constantly at fatigue duty up to this time from five (5) to eight (8) hours every day Sundays excepted." White troops were doing almost no fatigue work, and his regiment still had to take its share of picket duty. He acknowledged that official policy was that whites and blacks should do the same amount of labor but claimed that in actuality the work was much less evenly divided.[31]

In spite of his "three hours a day" claim on 7 July, by 24 July 1864 Kelley found that "Fatigue parties are at work night and day on the fortifications and will soon have them done." Perhaps not coincidentally, he also noted deaths were increasing in his regiment. "I have just put in a requisition for twelve coffins. How long do you suppose they will last the old Dr.? If he don't call on me for more bones next Sunday I shall be disappointed. The mortality in our Regt. beats anything I ever saw. They frequently drop dead in the streets, and in two or three instances have been found laying dead in the weeds some distance from camp. It is poor policy to take a regiment south in the spring of the year, especially a colored regt."[32]

News about the horrific conditions in Louisiana began to reach the North. In August a soldier wrote Abraham Lincoln directly, seeking relief. Nimrod Rowley was a private in the 20th U.S.C. I., which had been organized at Riker's Island, New York, in February 1864. By March the unit was assigned to the defenses around New Orleans. Rowley signed on wanting to fight the Rebels, but "Instead of the musket It is the spad and the Whelbarrow and the Axe cuting in one of the most horable swamps in Louisiana stinking and misery." Men were put to such hard labor, even if the day before they had been on the sick list. "By this treatment meney are thowen Back in sickness wich thay very seldom get over." Rowley was appalled by how many of his comrades had died in Louisiana. "We had when we Left New York over A thousand strong now we scarce rise Nine hundred." The food was inadequate, leaving the men in "A weak and starving Condition." Remember, he beseeched Lincoln, "we are men standing in Readiness to face thous vile traitors an Rebeles."[33]

All around Kelley and Rowley, men were dying at a great rate. In late October Gen. Daniel Ullmann, commander of the black regiments in Louisiana, commented in particular on the high mortality of the three regiments raised in St. Louis—the 62nd, 65th, and 67th. "The extraordinary amount of sickness and mortality in them presented such a remarkable contrast to the condition of the other colored Regiment, under my Command, that I considered it my duty to have the matter investigated . . . I desire to state in connection with them that I have never been able to procure sufficient Medical attendance, Surgeons, Hospital stewards & nurses, for these Regiments, nor has it ever been possible to obtain any but a very limited supply of vegetables."[34] Ullmann also noted that the heavy fatigue duty was taking its toll on the men. He did what he could to remedy the situation but without much success. He was particularly in a bind when it came to the labor assignments. He had jobs that needed to be accomplished, and as more and more men fell ill, the remainder had to do ever increasing amounts of work.[35]

One of the most damning descriptions of hospital life for black soldiers in Louisiana comes from Sgt. John Cajay, who had joined the 11th regiment of the U.S. Colored Heavy Artillery in Rhode Island. He wrote the *Weekly Anglo African* in September 1864 from a hospital at Fort Jackson, a river fort defending New Orleans. First, he wanted readers to know "what become of the monies raised by the Sanitary Commission for the relief of the poor colored soldiers. We do not receive one thing appropriated by your Commission. We get no dried apples, pickles, or potatoes, things we have not seen for two months, and if any comes to the fort by accident in boats they are all gobbled up by the grand functionaries who revel in the luxuries of life."[36]

Cajay also had harsh words for some of the hospital personnel. While many of the physicians and nurses were kind and attentive, other nurses were "low, mean men, and treat their patients shamefully." He was particularly critical of the hospital stewards, the men in charge of the medicine supply. "Even the hospital stewards who have the lives of the men in their hands, are not fit for the position they hold. Instead of attending to their business they both get intoxicated, neglect their business and get drunk and lay across the tables, and when the attendants ask for their medicines they tell them, "Oh! I do not feel well, wait until to-morrow morning." He went on to describe the inadequacy of the food and to lament that burials occurred without a minister presiding. "The mortality down here is very great; no less than two to three dying every day. We have lost seven men in two weeks . . . Out of one hundred and forty, one hundred and twelve remain, and if we stay in these swamps we will all die," Cajay concluded.[37]

Cajay's was not the only account of food intended for the African American troops being purloined by white officers. Col. Henry Frisbie despaired that officers in his regiment (the 92nd U.S.C.I.) stole food meant for enlisted men, many of whom were suffering from scurvy. "Some beef procured by a detail from my command . . . for the benefit of the troops who were suffering some with Scurvy was in my temporary absence taken by a cavalry officer who represented it to be by my order," he told his superior. "I am very sorry to say there are persons wearing the uniform of a United States officer who will not scruple to tell a falsehood to gain a petty advantage and use his uniform to deceive a "Poor nigger" and afterwards tell his smartness (shame) to his fellows and then with the air of a clown look around for applause."[38] Frisbie did not record the punishment, if any, meted out to this thief.

In the same month Duren Kelley boasted to his wife that "as the weather grows cooler my health continues to improve until I am about now as good as new. My old malady has entirely left me, and I believe that I am afflicted at the present time with nothing but laziness, and that you know is incurable as I shall apply no remedies." Kelley was also pleased to note that "the men are not dying off quite as fast as usual." In a month when privates were suffering from scurvy, and even men in the hospital did not receive adequate provisions, Kelley reported, "We are living first rate just now. Have plenty of nice wheat bread, corn cakes, flapjacks, pork bacon, fresh beef, *toungues* [sic], livers, hearts, &c." One wonders if Kelley was one of the smug clowns of whom Frisbie spoke. In any event, even he missed some foods. "We have no potatoes and can't afford to buy them of the settlers [sutlers?] at fifteen cents per pound. The paymaster hasn't arrived yet though we are looking for him every day."[39] If the potatoes, which would have cured scurvy, were too dear for the officers, they were even further out of reach for the privates who had also not been paid and lacked family support to supply the deficit.

The hospitals do not seem to have been up to the task of remedying the disease brought about by overwork, insufficient food, and exposure to typhoid and malaria. The shortage of physicians in Louisiana was particularly acute, as it must have been one of the least popular postings for physicians during the war. If before the war physicians were afraid to move to New Orleans, the conditions during the war made the place even less appealing. The resulting hospitals for black men, with few doctors and largely staffed by convalescents, were squalid. The Corps d'Afrique hospital in New Orleans was one of the worst. "The police of the hospital and grounds was bad," an army inspector noted. "The floors and bedding were dirty, and there seemed to be a lack of system and discipline."[40]

Louisiana was a mass graveyard for black troops in 1864 and 1865, with very little of the mortality inflicted by the enemy. Three regiments, with a total of 3,158 men enrolled, began the war at Benton Barracks but spent the bulk of the war in Louisiana. The three regiments lost four men to combat or accidents and 1,374 to disease. This 44 percent mortality from disease shocked those in higher command, who had thought the black man would not "suffer" in Louisiana.[41] Contemporary explanations included the usual cant about inadequate endurance or unusual weakness to certain diseases.

The real reasons were not lost on some observers at the time. The men were overworked in conditions of high heat, with inadequate food and clean water. The food lacked variety as well, and the presence of scurvy argues for pervasive malnutrition. By now army physicians were familiar with the scourge of measles among new troops. They also recognized that the black soldiers were camped in an environment widely heralded as poisonous. And it was, even if the "poison" was not the foul gases arising from rotting vegetable matter, as feared by contemporaries, but rather the clouds of mosquitoes that carried deadly malaria. Few Civil War camps were well enough policed to prevent contamination of water supplies by fecal material. The situation in Louisiana was exacerbated by the high water table, so that wells were inevitably in touch with fecal deposits. The presence of so many graves in the campground area of Port Hudson did not improve the quality and purity of the water, either.

Inadequate hospital care exacerbated this situation. Many of the nurses were convalescent soldiers. Aside from their limitations of skill and will, such men may also have served as active carriers of the diarrheal illnesses from which they were recovering. During the Spanish-American War, Walter Reed and his commission found that the convalescent nurse, untrained in hygiene and loaded with bacilli, was a principal vector of typhoid fever. The hospitals for black soldiers carried a reputation for general filthiness, and the convalescent nursing staff only added to the hospital germ pool.[42]

It is hard to determine the degree to which the white officers in the black regiments were to blame for the poor condition of their men's health. Like Duren Kelley, some no doubt saw the seasoning, whether by the measles or the climate, as an inevitable part of life in the army and Louisiana. The presence of scurvy is a direct indictment of the supply organization of the army, which was in easy reach of Union farms via the Mississippi River. Even the officers lacked potatoes, so the want of fresh vegetables was widespread. If it is true that officers stole meat from foraging expeditions intended for the rank and file, or diverted Sanitary Commission supplies intended for soldiers in the hospital, they were indeed despicable men. But there is not enough evidence here to convict more than a few of such behavior.

The main solution for all this suffering and misery was to end the war and muster the men out to their homes. This did not begin to happen in Louisiana until the fall of 1865, when the government finally conceded that it was time to reduce strength, as the South did look, at last, to be conquered. But some unfortunate men, signed up for three years in 1864, learned that they were not to go home yet. Instead, their regiments were moving to Texas. They could have had little idea of what awaited them there or how much worse their daily lives were about to become.

CHAPTER SEVEN

Death on the Rio Grande

Noah Davis had traveled many miles since those exciting days in the fall of 1863, when he joined the 8th U.S. Colored Infantry regiment at Camp William Penn, near Philadelphia. His regiment first moved to Hilton Head, South Carolina, and then marched into Florida, where it saw battle at Olustee. Back in Virginia, Noah had the satisfaction of watching Petersburg and Richmond fall to Union troops, and was at Appomattox when Robert E. Lee surrendered the Army of Northern Virginia on 9 April 1865. Like many soldiers he probably thought the war was nearly over and that he would soon return to civilian life. But Noah had signed up for three years, so when the order came for his regiment to board ships for Texas, he had little choice but to go. On 24 May 1865 his regiment left City Point, Virginia, for the three-week trip around the tip of Florida and through the Gulf of Mexico, bound for a little barrier island near Brownsville called Brazos Santiago.

Noah and his comrades were not in the best of health. They were tired of the army fare and welcomed the canned fruits occasionally provided by the Sanitary Commission. But there was little forage left in the war-ravaged countryside of Virginia, and many nights Noah went to bed hungry. It was getting harder to chew the hardtack crackers, for his gums were sore and his teeth loose. The voyage was particularly tedious, as there were no supplements to salt pork, hardtack, and coffee, morning, noon, and night. Noah was thirsty, too. The water ration on board was carefully doled out, and there was never enough to wash down the salty fare.

Arrival in Texas brought no relief. The food was the same. The water was worse. There was no fresh water on the island, so the men relied on a strange contraption that took salt out of the ocean water and delivered a barely drinkable brackish brew, steaming hot, to quench their thirst while laboring under the summer sun. As more and more regiments arrived, the water ration was cut and cut again. Finally, the men could bear it no more, and Noah's officers announced that they were moving about a hundred miles up the Rio Grande to the Ringgold Barracks. Noah agreed with his friends that the river water tasted nasty, but at least there was plenty of the muddy liquid to go around. Noah managed to stay on his feet until the regiment arrived at Ringgold but then could go no further. He entered the hospital there, diagnosed with scurvy.

By now Noah's gums bled at the lightest touch, his legs were swollen and ulcerated, and diarrhea plagued him. His doctors prescribed pulque, a drink made from the mashed leaves of the *Agave Americana* plant, which grew nearby in abundance. A cousin of a plant that is a key ingredient in tequila, *Agave Americana* was rumored to be a good treatment for scurvy. Noah drank it as ordered but felt no better. He grew weaker over time, and when his regiment was mustered out in October, Noah was transferred instead to the army's hospital in Brownsville. There he died on 7 December 1865.[1]

Noah died of a disease whose causation and cure were well understood in his own time. Why did he develop scurvy? And why, when hospitalized, did he receive such inadequate care that he died of the disease? His was not an isolated case. Rather, his story is the one surviving case report of a major outbreak of scurvy that is illustrative, if nothing else, of how not to feed and support an army. Hundreds of men died without seeing a single hostile action, and the army's response was to deny and cover up its officers' obvious malfeasance. The poor planning and execution of the deployment in Texas had its origins in inexperience and inadequate knowledge about the physical environment, but its taproot was a callous disregard for the 25,000 black men transferred there without adequate food or water.

The Invasion of Texas

"Where did the Civil War end?" a colleague asked me shortly after I came to Duke University. "Why, at Appomattox, when Lee surrendered, sometime in April 1865," I answered. "Wrong!" he crowed. "It ended when Joe Johnston surrendered his army to Sherman right here in our fair city, Durham, North Carolina." Johnston indeed made peace seventeen days after Appomattox, but hostilities between North and South still were not over. Confederate general Kirby Smith headed an army in Texas, and when General John Pope called on him to surrender on 9 May, he refused. Although the Confederate government had been in disarray since the fall of Richmond weeks earlier, many still hoped that its leading officials could reach Texas (perhaps with the elusive Confederate gold) and make a new stand. Even when Union troops captured President Jefferson Davis the next day, Smith held firm. The United States prepared to invade Texas, and on 11 May Union colonel Theodore Barrett led a few hundred men in a rash foray against Brownsville. They were met by a Confederate force five times their size and repulsed at a place called Palmetto Ranch. Of the Indiana troops, 114 lay dead, while only ten Confederates died in the action. This engagement was the last battle of the Civil War, and it was won by the Confederacy. Smith was jubilant, but even he could see that it was time to call it quits. On 26 May, he surrendered his command, and three days later Union troops marched into Brownsville unmolested. Other troops took up residence in Corpus Christi and Galveston.[2]

The situation in Texas caused continuing uneasiness in Washington. The state had not been successfully invaded during the war, with the exception of nips at its margins. It could become the staging ground for guerillas committed to keeping the Confederacy alive. Texas had been an independent country within recent memory. Who was to say it could not become so again? To its south lay Mexico, then a French colony ruled by Archduke Ferdinand Maximilian. The French emperor had installed Maximilian as emperor only two years before, with the support of one major faction in Mexican politics. The opposing side, led by Benito Juárez, continued to struggle against this foreign ruler and kept a stronghold in northern Mexico. The Confederate government had approved of Maximilian, hoping to win recognition and hence legitimacy from Mexico and France. Maximilian also condoned the shipment of cotton and other goods through the port of Matamoros, a town just across the Rio Grande from Brownsville, Texas, bypassing the Union blockade. The United States supported Juárez but stopped short of all-out military backing in fear of prompting a war with France. Thus a second reason for posting U.S. troops in Texas was to threaten the Mexican government with a show of force, while maintaining at least the partial fiction that the true goal was the pacification of this Confederate state.[3]

May of 1865 was not a great time for a commander in chief to announce that a large infusion of troops was needed for a new initiative on the Texas border. The country was sick of war and heartily ready to return to normalcy. A few white regiments and a single black one were already in Texas. Nine days before Smith's surrender, General U. S. Grant ordered Phil Sheridan to travel to Texas and prepare to take command of troops on the border there. The 25th Army Corps, composed entirely of black regiments, went to the area around Brownsville. Other white regiments were posted to San Antonio, Clarksville, Houston, and Galveston. General Godfrey Weitzel, who had led the troops that captured Richmond, commanded the 25th Army Corps, and was Sheridan's second in command for the Texas venture. Use of the black regiments made good political sense. Since these men had enlisted in 1863 or 1864, for a term of three years, they were available for duty in the postwar years. They had no political clout nor vocal folks back home to demand their discharge or even fair treatment. So Weitzel ordered his thirty-five black regiments to travel from Virginia to Texas in May and June 1865.[4] They did not all go willingly. When the 29th Connecticut Volunteers (Colored), stationed on board the ship *Demolay* on the James River, learned they were en route to Texas, there was an uprising. "Many were unruly, even threatening the lives of those who favored going to Texas whither we had been ordered for garrison duty," remembered on black sergeant. In addition to word of the unpopular posting, there were rumors that the ship was really going to Cuba, and the officers would sell the black men as slaves there. "Some of the gang were arrested for insubordination," and the mutiny quickly flickered out.[5] Ultimately, some 25,000 men took ship for Texas.

Life on Brazos Santiago

Their destination was a fort first built by General Zachary Taylor during the Mexican-American War. It perched on Brazos Santiago, a barrier island just opposite the mouth of the Rio Grande. A small contingent of

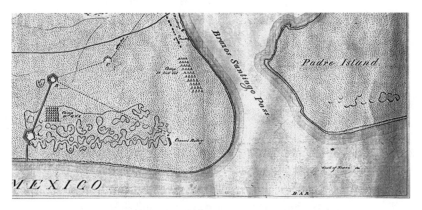

Map of Brazos Santiago Island, February 1865. "Made under direction Capt. P. C. Hains, USA, Act. Chief Engr., Dept of the Gulf." The small triangles represent clusters of soldiers' tents. At the bottom of the map is the Gulf of Mexico and at the top, the bay leading to the Rio Grande River and the Texas-Mexico border. This U.S. government map is archived at the Arnulfo L. Oliveira Memorial Library, Hunter Room Archives, University of Texas, Brownsville, Brownsville, Texas.

Union troops had returned to the fort in the summer of 1864, hoping to control the flow of Confederate cotton that avoided the Union blockade by crossing from Brownsville to the Mexican town of Matamoras, and from there to export from neutral Mexico's harbors. With a few hundred troops, life in the fort was just bearable. The thirty-five regiments that poured in during June 1865 quickly overwhelmed the fort's meager resources.

Brazos Santiago is not much more than a narrow sandbar with a port on the bay side. A description by a U.S. Sanitary Commission physician in 1865 reflects its barrenness. "This island is about five miles long, one half mile wide and raises but six or eight inches above the sea level except at the southern half which is covered with low, loose sand hills, entirely devoid of vegetation except a scattered saw weed which grows on the low hills and is the only species of verdure to be found on the island."[6] Just north of Brazos Santiago is South Padre Island, site today of coveted beach properties. But in 1865 all that sand was much less welcome. "This place is extremely hot, and is bordered with a low, sandy beach," wrote one despondent soldier to a northern black periodical. "There is scarcely a tree or shrub of any kind visible to the eye, not even a blade of grass. Every thing has a barren and desolated look."[7]

Compounding the influence of sand and sun was a shortage of water. "Most of the regiments have now landed," wrote private Samuel Smothers of the 45th U.S. Colored Infantry on 24 June. "The island is sandy and desolate-looking. We are put to considerable inconvenience about water: we have to use condensed water altogether, which is issued to us boiling hot, at the rate of three pints per day."[8] This "condensed water" came from a desalination machine, one of three or four that the army purchased to make seawater potable. These machines were essentially stills in which the vapor from boiling seawater was condensed and siphoned off. An apparatus for aerating the produced liquid supposedly improved the taste. But the men described it variously as brackish, sulfurous, or salty to the tongue. Not only did it taste bad, there was not enough of it. Estimates varied as to how much each man was allowed a day, from one cup to one pint, but all agreed that the shortage grew worse as more and more men arrived.[9] "The water we drank was condensed, and then at that rate we were allowed only one pint a day, and if one were not punctual, he would not get any at all. So you may judge our stay was not very pleasant," reported Private Charles Cole of the 51st U.S. Colored Infantry.[10] These men were performing heavy labor under the Texas sun and consuming salted meat on a regular basis.

"Our Colored troops were nearly perishing for water. The suffering was most intense," remembered Alexander Newton, a young black man serving as commissary sergeant for the 29th Connecticut Volunteers (Colored). "We paid ten cents a canteen for water and would have been willing to have paid fifty cents, or any price." The vendors were Mexicans, who bottled Rio Grande water and sold it at good profit. General Weitzel finally realized that keeping such a large contingent of men on the island was not feasible and ordered two divisions to march up the Rio Grande to create new encampments along the river. The march was hard, often through mud or water, but the men were grateful even for Rio Grande water. At one point some rushed to an overhanging bank, which collapsed and many drowned.[11] The river water was at best a moderate improvement. "This water is good when you have not tasted any for 50 hours," reported "Rufus" from Indianola, Texas, in mid-July. "But I would prefer Croton, if only for the appearance of the thing. The water of the Mississippi is crystal compared to it, and what is worse, is that all the sediment seems to be alive."[12]

Although the water deficit diminished as the men moved upstream, their diet did not get much better. "I am hungry enough to eat almost anything," wailed one soldier.[13] The diet on shipboard had been remarkable for the "total absence of fresh meat, potatoes, pickles, cabbage, beets, turnips &c &c. during the whole voyage of twenty-nine days," said a Sanitary Commission observer.[14] There had been no relief when the men arrived in Texas. "I have not seen a lemon, peach, apple or pear, nor corn enough over all that part of the country through which we have passed, to fatten a six months' pig," reported an army chaplain. "How the people live another writer must describe, and not I. It may be better farther in the interior, but few are willing to search for it."[15] The officers had canned fruits and vegetables in their mess, but none of it reached the privates' table. One source mentioned sweet potatoes, so perhaps there was some infusion of vitamins A and C, but not much.[16] Overall, the situation was growing steadily worse. Noah was hardly alone in the hospital. Another soldier, still on his feet, wrote to a newspaper, "Mr. Editor, to give you an idea of our sufferings, while here, our average loss is three men per day, for the want of food and water. Our poor fellows are still falling sick, going blind, and dying very fast."[17] The blindness was a symptom of vitamin A deficiency.

A Scurvy Epidemic

The men started falling so fast that they rapidly exhausted available hospital beds. There had been scurvy at various times and places during the Civil War but nothing to match the outbreak in Texas in the summer of 1865. Charles Smart wrote a graphic chapter on scurvy for the *Medical and Surgical History of the Civil War* in the 1880s. He noted that scurvy had a much greater prevalence among black than white troops, and that for the year from 1 July 1865 to 30 June 1866, the prevalence of scurvy was 141 cases per 1,000 men, with 2 percent of those cases proving fatal. The rates of disease in Texas were much higher. Surgeon Stacy Hemenway of the 41st U.S. Colored Infantry estimated that 60 to 80 percent of the black

troops were sick with scurvy in the summer of 1865.[18] If one takes a conservative estimate, say that 50 percent of the men were ill and only 2 percent of those died, that comes to around 2,500 scurvy deaths. The army's
official count said that 128 black soldiers died of scurvy during the summer and fall months of 1865, but this figure no doubt represents under-
reporting of cases and fatalities.[19] Hundreds of men fell ill in July, and
many of them were shipped to hospitals in New Orleans and elsewhere.

The hospital registers for the Brownsville Post Hospital survive. It is
unclear how representative these registers are and what percentage of the
hospitalized men in Texas ended up in this hospital. But some conclusions can be drawn from this data. The hospital did not open until 14 July
1865. Over the next two weeks, 842 patients entered the hospital; 46 percent of them had scurvy. Another 563 men arrived in August; 54 percent
had scurvy as an admitting diagnosis. There were only 32 deaths in July,
but over the next month 129 men died in the hospital. Combining these
six weeks, 35 deaths were from scurvy and 77 from diarrhea or dysentery.
Even though surgeon Ira Perry sent the case report of Noah Davis to the
surgeon general as an exemplar of scurvy, Noah's death is actually listed
in the register as being due to diarrhea. Scurvy causes diarrhea, and in severe cases a bloody diarrhea, one hallmark of dysentery. Gastrointestinal
complaints were common in Civil War camps and a common cause of
death. But given the mislabeling of Noah's case, it seems probable that
many of these diarrhea or dysentery deaths were, in fact, caused primarily by scurvy. The scurvy (and diarrhea) cases continued to be admitted
into the fall months, although by Christmas the epidemic appears to have
tapered off. This decrease reflected in part that regiments started going
home in October and November.[20]

Scurvy is not an all-or-nothing disease, like smallpox. Symptoms can
range from mild to fatal, depending on degree and length of deficiency
and probably on characteristics that vary among individuals. Many men
had scurvy during the summer of 1865 in Texas, but only a small proportion of them ended up in hospital. Some were not that sick; others were
cared for in camp because there were just not enough beds in the hospital. Not a few were buried without benefit of a hospital stay. Some regiments lost two or three men a day this way. They were buried in the Texas
sand, usually in unmarked graves. Garland H. White, chaplain to the 28th

Diagram showing the Prevalence of Scurvy among the White and Colored Troops of the United States during the War of the Rebellion

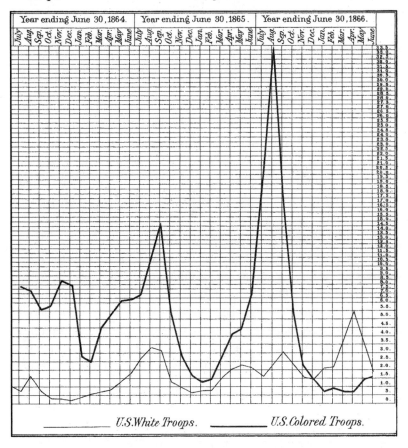

In this contemporary chart of scurvy cases, numbers are expressed as cases per 1,000 men. The peak number in August 1865 is 33.5. The denominator for the calculation would have included all black men still in uniform, not just those in Texas, diluting this result. From *MSHW*, 6:695.

U.S. Colored Infantry told the *Christian Recorder* in mid-September, "No set of men in any country ever suffered more severely than we in Texas. Death has made fearful gaps in every regiment. Going to the grave with the dead is as common to me as going to bed, for I also attend on such occasions in other regiments, rather than see men buried indecently."[21]

How It Happened

Why was scurvy such a problem in Texas during the summer of 1865? Did physicians and officers lack the knowledge to prevent it? Was their failure to bring enough fruit or vegetable provisions with them a sign of unbelievably poor planning? What had the Union army learned during the previous four years about preventing malnutrition among troops? Although my ultimate judgment is that this scurvy outbreak and its management reflected a mix of incompetence and indifference on the part of senior command, there are some mitigating factors that help explain why the 25th Army Corps (composed of all black units) found so much trouble in Texas.

The cause, prevention, and treatment of scurvy were well known in 1865. A century earlier, British naval surgeon James Lind had linked the disease to inadequate fruit in the diet and famously prevented its onset by the regular issue of lime juice. By the 1860s physicians were in full agreement that a diet with adequate fruits and/or vegetables was important in preventing scurvy, although they were also likely to call on a host of contributing causes when the disease broke out.[22] The Sanitary Commission's Dr. McDonald listed the absence of fresh meat, vegetables, and fruits as beginning the scurvy epidemic on the transport ships between Virginia and Texas, but he also concluded, "I think the cause of the scurvy was owing to the crowded condition of the boat, imperfect ventilation & insufficient exercise."[23] Others might have cited abrupt change of temperature, exposure to damp, inadequately ventilated tents, or excessive work in a hot sun—all typical explanations of disease or enervation in the mid-nineteenth century.

During the war the Sanitary Commission had clearly demonstrated that when scurvy did appear among the troops, food supplements could cure the problem. The Union lacked ample supplies of lime juice but did have an abundance of potatoes and onions, which seemed to reverse scorbutic symptoms admirably. And potatoes and onions could be shipped in barrels just about anywhere the train tracks reached. The commission also relied on pickled cucumbers, cabbage, and other vegetables. They might supply canned fruits and vegetables, although these were more expensive. The army came to count on the Sanitary Commission to step in when regular supplies were inadequate to stave off nutritional deficiencies.[24]

Army commanders also expected their soldiers to acquire fresh fruit and vegetables from local markets or suppliers. The official ration of salt meat, hard crackers, coffee, and sometimes beans was set artificially high, with the idea that commissary officials could sell the excess and use the funds to buy local goods. This was supposed to happen in hospitals as well, where the patient's regular ration could, in effect, be exchanged for the more delicate food needed in his weakened state. In addition, soldiers could buy extra food from sutlers, peddlers who were officially sanctioned in camp to sell the troops food, clothing, or other items. No one really expected a soldier to survive for months on the official army ration. Finally, it was accepted practice that soldiers should forage for food, whether in friendly country or hostile. Sometimes the owners of the animals or produce—or wooden fences, for that matter—would be paid for what was taken, but mostly soldiers confiscated what they needed from friend and foe alike.[25]

When scurvy occurred during the Civil War, it was usually due to unavoidable circumstances related to troop disposition and was fairly quickly resolved when the problem became apparent or the peculiar circumstances no longer applied. When Grant besieged Vicksburg, he was cut off from his supply train, and accessible produce from nearby farms was quickly exhausted. In the fall of 1863, Union troops were bottled up and starving in Chattanooga, as Rebel troops had seized the railroad lines leading from there to Nashville, the closest supply depot. When Sherman's men chased Joseph Johnston through north Georgia, the retreating Rebels left behind no food, and Sherman often moved too quickly for the tenuous railroad link back into Tennessee to keep his men fed. In all these cases, scurvy developed but resolved as supply was reestablished. In others, the Sanitary Commission stepped in quickly when scurvy appeared and prevented major epidemics.[26]

No one seems to have explained this history to General Weitzel or his subordinates, however. Some reflection might have revealed that the usual course of food acquisition was not going to work on the Texas border. In the first place, the previous year's events in the same area demonstrated that it was a prime spot for scurvy. The disease appeared in the fall of 1864, among the few regiments stationed at Brazos Santiago and on the ships that guarded the mouth of the Rio Grande.[27] "The health of the troops under my command is fair, although the want of fresh vegetables is felt in no small degree," wrote one colonel from Brazos Santiago in August 1864. "Two-thirds of the men in hospital are afflicted with the scurvy."[28] These were white troops from Illinois. Black troops stationed there fared even worse, for they suffered not only from scurvy and other diseases but faced callous physicians who scoffed at their suffering.[29]

It does not seem to have occurred to General Weitzel or his staff to stock up on fruits or vegetables on the way to Texas, even when their ships stopped in Mobile or New Orleans for refueling. The ships sailed past orange groves on their way up the Mississippi to New Orleans. With his men already showing signs of scurvy, the surgeon of the 22nd U.S. Colored Infantry noted with longing the "orange groves loaded down with golden fruit [which] delighted the eyes but did not satisfy the palate."[30] As it turned out there would be scant local produce to buy once the regiments arrived in Texas. Attempts to purchase fruits or vegetables from Mexican vendors fell through because of fighting among various guerilla groups, some of whom opposed selling produce to the Union troops. Officers had to rely on the canned products brought with them, as there was nothing else to find in Brownsville and environs. What little local forage could be found on the march went to white soldiers, as blacks were forbidden such activity.[31] Even the hospitals could not supply "anti-scorbutic" foods for the patients. The 25th Army Corps was stationed in a desert, and any relief would have to come from outside.

The men felt their isolation. One lieutenant with the 36th U.S. Colored Infantry wrote his cousin Gus in mid-July, expressing gratitude for a recent letter. "It had been a long while since I had heard from the civilised world, [here where] all we can see is sand and salt water and that is what we have to live on eat the sand and drink the water so you may imagine we are living high and sleeping in the cellar. It is called a very healthy place that is if we can only get plenty of fresh provisions vegetables which I do not see any prospect of our doing."[32] Another officer told his wife, "Brownsville is situated on the Rio Grande & is one of the most forsaken looking holes you ever saw."[33] A third soldier agreed, telling an African American newspaper that "a more God-forsaken spot does not [exist] in the wide world."[34] The trip from Texas to the East Coast by ship took at least two weeks, and faster communication across the war-ravaged South by train was not yet possible. The men were mostly isolated from the eyes of potential benefactors, and their officers were either too inept, powerless, or callous to remedy the situation.

Even hospitalization did not reverse the effects of scurvy for many men. Noah was under medical care from mid-July until his death in early December, with his last five weeks spent in the hospital at Brownsville. Hospital surgeons were supposed to use the hospital fund, generated by the value of the patients' unused regular ration, to buy foods specific to needs of the men in their care. But black men suffering from scurvy received the usual salt pork, coffee, and hardtack crackers as they did in camp. One man reported hospital patients gnawing hardtack with scurvy ravaged mouths, sending rivulets of blood across the white crackers. Dr. Macdonald of the Sanitary Commission heard that attempts to import fruits and vegetables from Mexico for the hospitals had failed. This may well have been true, but he missed seeing the larger corruption in physician behavior that left these patients so bereft.

Many surgeons were diverting money that should have been used to buy food for patients. Surgeons Charles Radmore, in charge of the Brownsville Hospital, W. A. Conover, chief medical officer of the Rio Grande, and probably others ordered "certain quantities of Brandy, Whiskey, Claret Wine, Champagne, Hostetters Bitters and Malt Liquors for the benefit of the Govt Service, almost none of which ever found its way to the sick in Hospital or elsewhere." These surgeons supplied medical care, for a fee, to private citizens and otherwise made money off hospital operations. According to a later investigation, "the Sick under Surg Radmore's charge were poorly supplied both with food and stimulants. Patients convalescing from Scurvy &c did not receive proper diet, so that in some parts of the command it was considered tantamount to a punishment or other great misfortune to be sent to this Hospital. These facts are notorious."[35] Investigation and correction of such malfeasance came too late for Noah Davis and many others.

Relief

From 1862 on, the Union army could count on the U.S. Sanitary Commission to bail it out when poor planning left its soldiers malnourished, inadequately clothed, lacking medicine, or in need of blankets. Fortunately for the soldiers in Texas, the Sanitary Commission still had an office in New Orleans in the summer of 1865, although operations were shutting down throughout the country. The war was, after all, over, and such an emergency relief organization hardly seemed necessary for peacetime life. Sanitary Commission agent Alexander McDonald was in New Orleans and heard about the hundreds of men pouring into local military hospitals, suffering from scurvy. After visiting the wards himself and hearing their stories, he took ship for Texas. What he found there shocked and appalled him. He wrote on 21 July, "Scurvy is prevalent to an alarming extent among the men and no anti-scorbutics can be obtained." He found 575 patients in the Brownsville hospitals, with two men dying every day. "At Brazos there are about three hundred and fifty patients and about four hundred have been sent to New Orleans since June 1st." The problem extended throughout the Texas camps. "All the troops here at Brazos are *suffering* for vegetable food and unless some is soon obtained we shall have an alarming number of cases of scurvy in the command."[36]

Sanitary Commission medical director Elisha Harris traveled to New Orleans to check out the situation. He too saw bed after bed filled with black men suffering from scurvy. He wrote to the home office, "The region there occupied by the 25th Corps is a *desert*. Without vegetable food the negro soldiers will die."[37] He pleaded with them to send a supply of potatoes and onions to the Texas coast. He called on the pride of tradition, saying "It is conceded by all military authorities in the Mississippi Valley that the Sany Comm has done what all the military authorities of the armies here could not do in supplying vegetables."[38] Instead, the plan was to close the New Orleans office and ignore the problems in Texas. The war was over. A week later, Harris begged again. "The Sanitary Commission will never do better service to the colored soldier & the nation than by sending vegetables to the shore of Texas ("Coals to Newcastle", you might think,) and *then* recording and telling of how & why this was done."[39]

This last statement may have done the trick, for the Sanitary Commission did send relief supplies, including pickles, potatoes, and onions, to Texas. The thrust of Harris's last phrase was that the story of the situation in Texas would give the army, and particularly the army's medical department, yet another black eye. Relations between the Sanitary Commission and army surgeon general Joseph Barnes were frosty at best. Barnes was the beneficiary of a cabal, headed by Secretary of War Stanton, which had impeached Surgeon General William Hammond earlier in the war; Hammond had been a strong friend of the Sanitary Commission. The commission's very existence pointed to inadequacies in the War Department, and although individual physicians were often cordial to USSC work and praised their efforts, the attitude at the highest levels was one of hostility.[40] Barnes had just issued an order via telegraph in June 1865, which commanded army physicians not to talk to USSC people, especially those physicians who were gathering information in order to write a "Sanitary History" of the war. Barnes's justification for such a gag rule was that the army was going to write a history of its own, and the Sanitary Commission's agents would merely duplicate that, while pestering busy physicians and keeping them from their work.[41] So now, in the summer of 1865, the plea was out for the Sanitary Commission to once again save the army's reputation.

One can understand the reluctance to step in one more time, but as Harris pointed out in his letter, it would give them one more example of Sanitary Commission heroism in the face of Army incompetence. "The *crime* of causing all this suffering rests solely upon the head of the Hon. Secy. of War & on his alter ego in the responsibilities of the Medical Bureau."[42] With some glee Harris anticipated showing the world that "the War Office is blind, ignorant, and heartless." And he fully planned to continue gathering information for the sanitary history. "We shall proceed," he wrote, as if Surgeon General Barnes "were our best friend & the Bureau a promoter of science & humane improvement."[43] Alexander McDonald was there in August to see the Sanitary Commission supplies unloaded and delivered to the hungry men.

No doubt the Sanitary Commission fruits and vegetables made a difference, and McDonald reported that the army also had altered their standard ration to reflect the local circumstances. I have been unable to determine exactly how much of what kind of food arrived. McDonald signed off on his report concerning the Texas fiasco on 30 October, implying that the crisis was over.[44] The Brownsville Hospital ledger showed men still being admitted for scurvy in November, however, and Noah Davis died of the disease in December. So the supplies helped but did not entirely remedy the situation. Most important for the men in Texas were the orders that started coming in October and November. They were being mustered out, and the best cure for scurvy would be found in the markets back home. The political instability that had prompted the ill-fated expedition in the first place had quieted, and the army left only a few regiments to guard the now quiet front.[45]

The Official Response

The army's response to the crisis in Texas was to deny that it was happening and to prosecute anyone who claimed otherwise. The first such reaction came in response to the letter from M. R. Williams published in the *Christian Recorder* on 30 June 1865.[46] Williams described the inadequate food and water, callous physicians, and brutal working conditions, including marches through mud or knee-high water. Two months later a letter from William R. Miller, sergeant major in the 22nd U.S. Colored Infantry, rebutted Williams's claims. Miller scoffed at inconsistencies in the first letter and all but said that the reason Williams and comrades had trouble with doctors was the soldiers' malingering. He signed the letter "for my race and justice."[47] It is impossible to tell from this correspondence whether Miller was white or black. What is obvious from the computerized database of soldiers and sailors in the Civil War is that Miller existed, and Williams did not, at least there was nobody by that name in the company and regiment given in his letter.[48] Still, Williams's claims are supported by multiple other independent sources. It is clear that Miller was denying events that had actually happened, perhaps at the insistence of his commanding officers.

There the matter probably would have ended, as the distribution of the *Christian Recorder*, an African Methodist Episcopal journal, probably did not extend far into the ranks of white officers. But in early September a newspaper in Galveston reprinted the initial Williams letter, and the army hierarchy bristled. The assistant adjutant general for Texas ordered Llew-ellyn Haskell to find this Williams fellow. Haskell reported back that "there is no man of that name either in the company or regiment named, nor has there ever been—there is no Williams in Co. 'E' that is thought sufficiently educated to write such a letter." Haskell called the letter a "scandalously false statement" and mused that the author probably used a pseudonym "for the reason that he so well knew the falseness of his statements at time of writing them." This report found its way to General Weitzel, who raged, "Statements contained in this document are evidently false. The author of it should be immediately arrested and tried for viola-tion of the 99th article of war."[49] One man literate enough in 1865 to have written the letter was Alexander Newton, a black sergeant in the Con-necticut African regiment, who later trained to be an African Methodist Episcopal minister and wrote his memoirs, including a critical descrip-tion of the unfortunate events in Texas. Yet he did not fall under suspicion of having written the letter at the time of the army's investigation.[50]

Another whistle-blower was not so fortunate as to be hidden by ano-nymity. Surgeon George Potts signed his name to his complaint. Potts had joined the army in 1864, after service as a physician for the East Indies Company and at the British consulate in Siam. Appointed surgeon of the 23rd U.S. Colored Infantry, he first came to grief in the spring of 1865, when he was court-martialed for behavior surrounding the autopsy of a black soldier. What Potts argued was proper procedure for preserving bod-ily remains for study, others saw as the desecration of a body. The court convicted Potts, but Surgeon General Barnes must have agreed with the doctor, because Barnes reversed the decision and reinstated Potts as a sur-geon. In a moment of signal bad judgment, Barnes sent Potts back to his old unit, where he clearly had already earned the enmity of the command-ing officers.[51]

Potts was not one to sit quietly in the face of rank injustice. He was sure that if Surgeon General Barnes knew the true state of affairs in Texas, Barnes would take appropriate action. "It is impossible for one situated as I am to form any correct approximate of the enormous per centage of the sick throughout this Corps," he wrote Barnes on 20 July. "Scurvy how-ever prevails every where. And whoever was in charge of the Medical De-partment of this expedition has shown an utter disregard of Human life."

Potts revealed his frustration at inadequate supplies and his awareness of the prior summer's scurvy outbreaks. "No precautions[,] no provision whatever appears to have been made for ready access to obtain supplies of anti-scorbutics and up to the present time none can be obtained notwithstanding that past experience of the ravages of this disease amongst our Troops operating in this part of the Country ought to have been sufficient fore-warn and consequently to fore-arm against such dreadful consequences." Potts was well aware that petitioning Barnes could bring retribution on his head. "I would humbly present these facts before the Surgeon General USA knowing that he is not aware of the true state of affairs, and aware that such a sweeping accusation exposes me to the vengeance of all concerned and would solicit his protection while I fearlessly discharge this painful duty."[52]

Potts not only lambasted the officers who had allowed scurvy to strike their men but also particularly criticized his fellow physicians. They were lazy and covered up for each other. "The consequence is that the Sick in the Hospitals complain that they are not visited regularly every day by any Medical Officer. Some of the sick describe the visits as simply a look into the ward and state, that occasionally a few patients are prescribed for. That when prescriptions are made the medicine is hardly ever administered." As a consequence, he reported, sick men resisted hospitalization as long as possible, preferring the care of comrades to the indifference of the hospital staff. Such a situation could bring only opprobrium on the medical profession.[53]

The endorsements on the back of this letter indicated that it was directed back to Texas for investigation. General Weitzel's response, in spite of a personal plea from Potts, was to place Potts under arrest. He was prosecuted by a board of the very surgeons he had condemned as corrupt. Potts described his opponents as vengeful and full of hatred toward him. Their investigation blamed the presence of scurvy in Pott's regiment on his poor management, ignoring the fact that every black regiment in Texas was suffering mightily from a want of vegetables and fruit. Potts made a convenient scapegoat for the whole mess, however, and on 18 October he was found guilty of all charges and cashiered from the army with none of the back pay due to him.[54]

There the matter rested until after Christmas. Sometime in the interim, surgeon C. B. White ceased being acting medical director of Texas, and Edward P. Vollum took over. The latter must have reopened the investigation because by January he had formed his own opinion about Potts's case. Vollum found that "Dr. Potts appears to have been regarded by some as insane and by others as vicious and naturally troublesome, at any rate his statements and complaints, thrown out in an irregular and voluminous manner, have not been regarded as reliable by his superiors in this Department." But Vollum did not stop there. "Yet," he continued, "there is a ray of truth in his letters. The 25th Army Corps has been remarkably unhealthy since its arrival in Texas, exceeding in amount of deaths any other organization in the Army at any time." Vollum had not been there and was able to look at the story with some objectivity. "The diseases seem to have been chiefly of a scorbutic character and one would suppose that with the ample resources of the Medical and Commissary Departments that preventive measures might have been adopted, which, as far as I can learn, were not even attempted." Vollum spoke diplomatically, but his anger showed through. "I therefore cannot avoid the conclusion, that the Medical Officers of the Corps, were too inexperienced— the mildest view of the case,—to represent the Medical Department in so important an event, as the outfitting and shipment of 25000 troops from Johns river to Rio Grande." "Inexperienced" was the kindest label possible. Vollum concluded, "The Medical Affairs on the Rio Grande have improved very much of late, and I believe that the above complaints do not apply now to that section in any way but that there has been a great amount of ignorance, indifference or mismanagement in the Medical Affairs on the Rio Grande heretofore there is no doubt in my mind whatever."[55]

It seems Vollum went further, for he stirred up Surgeon E. M. Pease to investigate the charges of hospital mismanagement in Brownsville. Potts had labeled Pease as one of his particular enemies, so Pease may have been in the dicey situation of finding someone to blame while keeping his own name out of the case. In any event, Pease laid the blame on several surgeons who had already been mustered out of the service and on Charles Radmore. Pease indicted Radmore for stealing hospital funds and buying liquor with them. He concluded, "The list of accusations might be made considerably longer, but I have preferred to touch only the more obvious facts and such as are most susceptible of proof." Pease laments that "every word I have written could be full proven were the necessary witnesses—many of whom have left the State . . . to testify." Conveniently, many of the men affected were scattered and unavailable to offer testimony. Piously, Pease went on, "Fraud and corruption seem to have been common principles of action[,] to make money by whatever means the great end and object of pursuit."[56]

Pease was hopeful that the whole conversation could now be at end. "But whether it is now advisable to investigate these things in the face of so much difficulty and opposition as is sure to be encountered I leave entirely to your judgment and decision. Personally I would prefer to have the whole subject dropped forever, but I could not do my duty and remain entirely silent."[57] It seems that Pease got his wish. While Vollum wrote on the back of the Pease letter that he recommended further investigation, there is no further discussion in Radmore's file of his offenses. Radmore himself stayed in the army as a surgeon until March 1867.[58] Pease, who began his career as a surgeon to the black regiments by complaining about the preference in rank shown to Alexander Augusta, left the army to serve as a missionary in the Marshall Islands.[59]

Retrospective

The first published account of the Texas scurvy epidemic appeared in the *Chicago Medical Examiner* and was written by Stacy Hemenway, surgeon to the 41st U.S. Colored Infantry. Hemenway described the high case and mortality rates from scurvy in Texas but assigned no blame for the disaster. Perhaps he had learned from George Potts and kept his opinions about causation to himself. Also, by this point, Hemenway was the "late surgeon" to the 41st regiment; he was now in private practice in Chicago.[60] Hemenway reported that 11 percent of the men treated in hospital died, and many more were sick in quarters. Men poured into the eighty-bed hospital so quickly that within a week he had "no less than 500 patients," and "many [were] lying around the hospital buildings, without shelter, in the open air, upon the sand" with only limited supplies of water under the bright Texas sky. As fast as Hemenway could get the men loaded on ships for New Orleans, more arrived in similar desperate condition.[61]

Hemenway does not attribute the soldiers' illness to innate black characteristics. Instead, he reports that he recognized scurvy among the men before the regiment had left Virginia, and their condition was only compounded by the sea voyage from Virginia to Texas. The men were crowded into spaces between decks that "were commonly dark and illy ventilated, thus constituting the chief exciting and immediate cause." Their diet in Virginia had been inadequate, and often the issued rations were spoiled, with "wormy hard-bread and a damaged condition of beans, rice, and pork. The supply of fresh vegetables were very meagre." Things only got worse in Texas, where the men were assigned to excessive fatigue duty, housing was inadequate, and the water impure. It is clear from Hemenway's account that the men died because of remediable, external causes.[62]

Surgeon Sanford Hunt, who had served in the army and was also a major figure in Sanitary Commission circles, was less diplomatic in his assessment of events in Texas. In an article on scurvy during the war, Hunt explained the etiology and pathophysiology of scurvy and described various outbreaks. He saved his strongest vitriol for condemning the army for what happened in Texas. After recalling the lessons about scurvy hard won during the Crimean war, Hunt fumed "Even the . . . hard teachings of four years of campaigning upon our own soil were not enough to induce the government to prevent one of the most remarkable and extensive prevalences of scurvy that the world has known, from occurring in the Rio Grande Expedition after the war had closed." Hunt considered the outbreak inexcusable. "This latter instance is one for which there was no sufficient excuse. During the war the existence of scurvy had been so popularly recognized that throughout the whole north charitable associations were active in forwarding 'antiscorbutics' to the storehouses of the Sanitary Commission." What every housewife in Ohio knew, however, "produced a hardly perceptible influence upon the Commissariat."[63] Hunt shared, obviously, the animosity of the Sanitary Commission's leaders toward the army hierarchy, but his judgment rings true.[64]

Charles Smart, writing the official medical history of the war in the 1880s, did not share in Hunt's opinion. In his article on scurvy, he concluded that "the very great prevalence at the period mentioned [Texas, summer 1865] must therefore be attributed to a deficient dietary operating on the system of a race having perhaps a special predisposition to be harmfully affected by the deficiency." Smart did acknowledge that he had no special reports to draw on from the summer of 1865, so he made his assumptions about racial susceptibility in spite of an absence of data. Smart was more on target in his broader conclusions about scurvy and war. "It cannot be said that the history of scurvy in our armies has added much to our knowledge. It shows how readily the disease may be controlled by the adoption of appropriate measures; but this had already been repeatedly illustrated," Smart argued. "It shows that although the law may provide adequate means for the prevention of the disease, the desired and anticipated results may not always follow on account of difficulty in procuring or transporting the supplies needful for large bodies of men under the changeful conditions of active military service. Perhaps this is its most instructive lesson." With prescience, Smart concluded, "From it may be foreseen the occasional appearance of the disease in time of war, unless the anti-scorbutic principle be meanwhile obtained in a form in which its issue to the troops will be more frequently practicable than when associated with fresh beef of the hoof and potatoes in barrels."[65]

It is hard to view the epidemic of scurvy among black troops in Texas as anything less than the product of gross negligence and incompetence on the part of General Weitzel, the officer staff, and the physicians of the 25th Army Corps. As Hunt pointed out, the "hard teachings of four years of campaigning" had demonstrated the importance of an "anti-scorbutic" diet to even the densest of officers. And even if they had missed it before the summer of 1865, the occurrence of hundreds of cases of scurvy should have driven the lesson home. Still, there was no response. Many physicians ignored their duties, made money off the hospitals, and enjoyed the whiskey ration to the detriment of their patients, left gnawing hard crackers with bloody gums. Those who attempted to garner attention to the plight of black soldiers were in turn persecuted. By the time anyone in the army took the problem seriously, most men had been discharged and the culprits were home free.

It is one thing for deaths to occur in wartime; then troops expect to be in harm's way. But the summer of 1865 was a time of peace, with only small aftershocks of war, and there is no reason that young men should have died at all, had they been well housed and provisioned. By the standards and knowledge of the time, the deaths were entirely preventable. As

Elisha Harris wrote about the sea voyage from Virginia to Texas, "half dozen potatoes and onions to each man would have been enough to stave off illness."[66] Not much more would have kept them out of trouble all summer, and Noah Davis would have gone home in October with the rest of his regiment. The many factors that led to high black mortality from disease during the war came together in that deadly summer. The officers and physicians were at best incompetent and at worst corrupt and callous. The events happened far out of common view, and even the gaze and concern of the Sanitary Commission had trouble seeing the problem. There was little media attention, although the fact that the one published letter caused such a stir indicates that more exposure might have made a difference. Ira Perry, a surgeon at the Brownsville Hospital, did numerous autopsies on the scurvy patients, and his account of Noah Davis is preserved in the medical history of the war. It is too bad that along with the detailed description of internal organs, he and his colleagues could not see the injustice that had brought that body to the dissecting table.

CHAPTER EIGHT

Telling the Story

The most curious feature of postwar accounts of the black soldier's body is the missing perspective of the sympathetic physician. Ira Russell published two papers based on his experience at Benton Barracks, but their critique of the overall treatment of the black soldier is fairly mild and makes no direct charges of abuse. Instead, the point of view that the black soldier was inherently inferior was allowed to stand and was even codified by anthropometric statistics that emerged from the war. Sanford Hunt published Russell's autopsy data, and he cast it in terms expressive of the innate shortcomings of the African soldier. By the 1890s many had forgotten that the black man served at all in the Union armies, and those who did acknowledge the black presence in war did so only to draw conclusions about the inability of the black body to cope with the strains of army and, by extension, independent life.

Russell's paper on pneumonia at the Benton Barracks begins by noting that there were six regiments of black troops recruited there and that in the winter of 1863–1864 pneumonia was particularly prevalent among them.[1] Even in the midst of such misery, Russell maintained his scientific perspective. "I watched the progress of this disease with much interest and care. Autopsies were frequently made, and the morbid appearances carefully noted. I have endeavored to ascertain . . . the causes that operated in producing the disease, and the most effectual mode of combating it."[2] In explaining this wave of illness, Russell noted that the winter was very cold, "the like of which had never been experienced by that important personage, the 'oldest inhabitant.'"[3] The barracks were next described, in calmer language than he generally used in his letters to the Sanitary Commission. "One hundred men were crowded into rooms originally meant but for fifty, necessarily rendering the air very impure; and this evil was rendered greater by the faulty construction of the barracks and imperfect ventilation."[4] Finally, some sort of epidemic influence was at work, for hospital attendants who lived in more comfortable surroundings also suffered from the disease.

It is worth noting that the strongest conclusions drawn in the paper are not ones about which specific data are offered. Russell and his physician colleagues found that a supportive, stimulant plan of treatment, which included tonics, alcohol, and a nutritious diet worked better than heroic therapies such as bloodletting, antimony, and mercury. But there was no attempt here to test one mode of therapy against another. The many postmortems done generated much data on factors such as the weight of the lungs, degree of "hepatization" (a measure of severity), and the presence of effusions. He also gathered data on length of hospitalization, yet he does not comment on the practical importance of this information.

Russell does draw some general conclusions about "the characteristics of this disease as it affected different races and classes of persons." He did not believe that black men were inherently liable to pulmonary disease, as was widely argued. Russell had the opportunity to compare people of roughly the same socioeconomic class—the contraband turned soldier and the poor white refugee. He found that "pneumonia was prevalent among them all. The attacks were not as frequent among the white soldiers, nor were they as fatal. Among the other classes mentioned little difference was observed, except the greater frequency of the disease, and greater mortality among the white refugees."[5] But Russell goes on to emphasize that more pleuritic adhesions and scarring, indicative of old pulmonary disease, was found among blacks than whites (class not specified). Still, those same autopsies showed much less sign of tuberculosis than prevailing opinion would predict. Finally, the healthy black lung weighed four ounces less on average than the white, a fact that might explain, Russell posits, the "inability of the negro to endure forced marches with equal facility with the white soldier, as has been frequently shown during the war."[6]

Meningitis was also rampant among the troops at Benton Barracks during those fatal early months of 1864. Russell reported that some fifty cases appeared among the black troops, and half of those patients died. Then it spread to the white troops, and it perhaps surprised the physician that "the symptoms and progress of the disease [did] not differ . . . materially from that among the negroes."[7] Russell believed this disease was caused by "local miasmatic influences" and argued for use of opium and quinine in its treatment.

Although Russell survived the war by twenty-three years and published articles on insanity well into the 1880s, he never again returned to the subject of his black patients during the war.[8] The other physician sympathetic to the health of the black troops during the war likewise published a brief article and then abandoned the topic. Stacy Hemenway was surgeon to the 41st Regiment, USCT, and traveled with them to Texas in the summer of 1865. There he witnessed the scurvy outbreak, which afflicted anywhere from one-half to three-quarters of the men in his regiment. Hemenway placed the blame squarely on remediable factors such as diet, crowding, poor ventilation on shipboard, and impure water. He says nothing in his paper about any innate black susceptibility to scurvy.[9]

That is not the story promulgated by official sources, and other commonly cited summary accounts of the black man under wartime conditions. Instead, the black's innate weakness, defined in a variety of ways, is blamed for his tendency to die so readily from disease. On 17 March 1866, Provost Marshal Gen. James B. Fry sent his report to Edwin M. Stanton, Secretary of War. Fry's office had multiple tasks, which included keeping the statistics of the army. In the section of his report on casualty statistics, Fry acknowledged that "in the casualties among the colored troops the most striking circumstance is the enormous proportion of deaths by disease. The ratio is no less than 141.39 per thousand, while the . . . general [white] volunteer ratio is 59.22."[10]

He found this particularly surprising, since the black troops saw so little of "the hardships of field service proper" and suffered slightly less than half the mortality rate from battle wounds recorded for white troops. Fry found his explanation within the black body. The data, he argued, "seems to indicate that the negro, in the condition in which the war found him, was less able than the white to endure the exposures and annoyances of military service." That such a large proportion of the black troops were on the sick list at any given time was evidence of a lack of stamina. Fry thought he understood its origin. "It is merely suggested that it is moral rather than physical; that the great susceptibility of the colored man to disease rose from lack of heart, hope, and mental activity, and that a higher moral and intellectual culture would diminish the defect." Fry closed by stating that "this view is supported by the opinions of surgeons of boards of enrollment on the abstract question of the physical fitness of the colored men examined by them."[11]

Other authors uncritically echoed Fry's conclusions. Edward Dunster, who wrote a chapter on the comparative mortality of armies for the Sanitary Commission's memoir of the war, made no effort to explore the detailed reasons for the black and white mortality differences. After concluding that "these figures appear to indicate pretty conclusively that the negro, as he was found in our armies, was less capable than the white man of enduring the fatigues and hardships, and of withstanding the influences of disease incident to army life," Dunster explicitly agreed with Fry's conclusions about the black soldier's psychological weakness.[12] In the same volume, Roberts Bartholow echoed his own comments published earlier in a guide to the physical examination of recruits. Although Bartholow served in a general hospital in Nashville during the war, there is no evidence that he had extensive contact with the black soldier as a patient. He left the war with the same preconceptions with which he entered it: "The Negro soldier is, unquestioningly, less enduring than the white; less active, vigilant, and enterprising, and more given to malingering."[13] Neither Bartholow nor Dunster seems to have read Russell's article on pneumonia among the black troops or to have benefited from its insights regarding black mortality rates, even though Russell's paper appeared within the same collection of essays.

General Fry's comment that the enrollment surgeons who examined recruits agreed that the black man was inferior is peculiar, given that his own office would publish just the opposite claim nine years later. On 1 May 1865 the Provost-Marshal-General's Bureau sent a questionnaire to the examining physicians serving in the Union states and territories, asking their opinions on multiple issues. These were the men who examined fifty to one hundred men a day at recruitment stations, determining their fitness for service.[14] The physicians were guided by a manual written by Bartholow, which was adopted by the U.S. surgeon general and provided for their use.[15] In the time between 1 May 1865 and the following March 1866, when Fry wrote his report, the responses to the questionnaire had arrived, so it is odd that he so misrepresented their content regarding the black soldier.

Results from the questionnaire appeared in 1875 as part of a massive volume on the physical description of the recruit. One of the last queries on the list asked the physician's "experience as to the physical qualifications of the colored race for military service."[16] Some 188 questionnaires left the Provost-Marshal-General's office; 115 of those returned. Completion rates were best from the New England states (100%) and less so from the mid-Atlantic states (59%) and the Midwest (51%), which may bias the findings.[17] Abolitionist feeling ran strongest in New England, and with it perhaps a greater likelihood to look kindly upon the black recruit. The introduction summarized the physicians' findings as follows: "In reply to the inquiry as to the fitness of the negro for military service, a want of opportunity for observation has generally been alleged; but so far as the experience of the writers extended, it is noticeable that they all seem to speak with admiration of the physical proportions of the blacks who came before them."[18]

A closer reading of the responses recorded in this volume indicates less universal approbation of the black body. Only about 66 percent of the respondents ranked the black recruit as either the equal or the superior of the white man. Another 18 percent said they had insufficient grounds to judge the matter (and did not) and 16 percent found the black to be inferior. Since the question was about fitness for military service, a positive answer was not always flattering to the black man. The physician might in the same paragraph praise his physical strength while denigrating his intellect. The decreased intellectual powers did not detract from the black man's capacity for the soldier's role; in fact, it made him more amenable to the obedience and subservience expected of the private toward his officers.

The efficient examining physician had assistants who had the recruits undress, so he met them naked. This unequal meeting may have promoted the physicians' tendency to see all the men as animals judged in the marketplace, especially the black men who were, after all, being sold in that manner in the southern states. Physicians sometimes praised the black recruit as if he were a particularly fine animal specimen. A physician in Lawrence, Massachusetts, enthused about one recruit, "a man of prodigious muscular strength, a very Hercules, whose thewes and sinews would have done credit to a horse."[19] An Elmira, New York, examiner found that in the black man, "the muscle is more full and distinct, standing out with greater prominence, but they are more slow and heavy in their motions," lacking the mental capacity of the white race.[20] One New Jersey commentator agreed that "the colored race, physically, are well developed, muscular, and strong. Their organization denotes the possession of brute force rather than intellectual pre-eminence." His description continued in language that reinforced the animal analogy. "The facial angle is smaller than that of the white man, with prominence of the lower jaw, and with large muscles for mastication. The shoulders are massive, with powerful muscles attached to the superior extremities." This brutish appearance extended to the genital region. "The buttocks are flattened laterally and are prominent posteriorly, with the fissure between them sunk deeper and more compressed than that of the whites; in this the anus is deeply sunk, and is mostly free from disease. The penis is large and long, and not often scarred with chancres." He concluded that such beings would be well able to endure long marches and the duties of manual labor.[21] The most caustic of examining physicians, a doctor in Davenport, Iowa, concluded that the black man "has some capacity, physically considered, for military service, there cannot be a doubt; neither is there a doubt about the usefulness of the horse when subject to intelligent training."[22]

Even when these physicians were overtly racist in their judgment of the black man's cognitive ability, few saw reason to doubt his physical hardiness. They do appear to have believed that men of mixed race were weaker than those with "pure black" heritage. "I am convinced," said one, "that, as a general rule, any considerable admixture of white blood deteriorates the physique, impairs the powers of endurance, and almost always introduces a scrofulous taint."[23] An Ohio colleague agreed, reporting "that the mulatto and all varieties of mixture of black and white blood have degenerated physically, being very often found with tuberculosis and other manifestations of imperfect organism."[24]

The persistence of this opinion, especially among examiners who claimed to have seen few recruits of African descent, is puzzling. It is possible that free blacks, many of them coming from crowded, impoverished lodgings within the cities of the north, were more exposed to tuberculosis than whites. And perhaps more northern blacks were lighter skinned. But neither seems particularly likely. Consultation of Bartholow's manual for examining officers offers the best explanation for this repeated connection between skin tone and tuberculosis. In it he says, "Few negroes having admixture with white blood are free from scrofula." Particularly in cold latitudes, the mulatto "falls prey to scrofula, loses the power of reproduction, and becomes extinct in a few years."[25] Although some examiners reported that they had turned away recruits because of signs of tuberculosis, phthisis, or scrofula, most who expressed this opinion of disease prevalence among mixed-race candidates were merely parroting what they had read. Some of them may have been aware of the writings of Nott and others on the degenerating influence of racial mixture, but it is far more likely that they were affected by this manual that the government had put in their hands and ordered them to follow. The influence of the Bartholow manual can be seen indirectly in the statement of a brave physician from Hollidaysburgh, Pennsylvania, who found more tuberculosis among whites than blacks, so that his "experience has led me to differ from high medical authority."[26]

Altogether the impact of the Baxter volume was to praise the black man as strong and healthy when he entered the service. The black man had similar chest circumference and expansion when compared to the white recruit, both signs of normal pulmonary health; no other markers of physical inferiority were found on measurement. How could this jibe with the known statistics of black mortality in the war? Far from surpassing the black man in health, a higher percentage of whites were rejected for health reasons than blacks. Chart 22 of the report shows that while only 66 percent of white applicants passed the examination, 75 percent of blacks made it through.[27]

Some caveats should be kept in mind in approaching these numbers. Some 180,000 black men served in the northern armies; only 25,828 were included in this volume's statistics.[28] So most black recruits did not see one of these official examining physicians. Most of the ex-slaves who entered the Union ranks were not seen by a physician, and many of them were less healthy than the free black volunteer. Further, much of the data gathered in the first two years of the war was lost. The white recruits seen in 1863 – 65 were less likely to be as healthy or desirable as the men who enlisted early in the war, and many of them were draftees or substitutes. If all whites and all blacks were compared for conditions that should result in rejection for service, the numbers would probably be quite different. Finally, it should be remembered that the examining physicians saw the black body only for a few minutes; these physicians had no opportunities to see how the black man did in fact endure the hazards of camp life.

One of the most influential articles about the health of black soldiers was Sanford Hunt's piece that appeared in the *Anthropological Review* in 1869. It is not at all clear why Hunt wrote this article when Russell had by far the greater experience and had done most of the research. Hunt used Russell's data and quotes extensively from his reports to the Sanitary Commission, but Hunt's approach to the black soldier was far different from Russell's indignant sympathy. Hunt was more distant; there is little in his tone to betray personal familiarity with the men who were the subject of his discourse. Hunt was accordingly more prone to spout stereotypic sketches of the black temperament and body. For instance, the black was docile and passive. "In no instance did he assume leadership, in no instance did he organise to strike a blow for his own liberty."[29] Hunt lauded the "well known imitative faculty of the negro, together with his natural fondness for rhythmical movement" as generating great aptitude for drill. The black soldier quickly fell into patterns of obedience, although his level of cleanliness differed from regiment to regiment.[30]

Like Bartholow, Hunt commented on the "large, flat, inelastic foot of the negro," which some had felt would prevent long marches. This fear had been obviated by experience; the negro marched as well as any soldier. In fact, "his large joints and projecting apophyses of bone gave a strong leverage to the muscles attached to or inserted in them." Hunt cited various opinions about the black soldier's capacity for enduring fatigue and hunger. Some felt he did well at this, others that he "is, *at present,* too animal to have moral courage or endurance." Hunt concluded that if the troops were well-fed, and an officer worked to keep their spirits up, the black troops performed well.[31]

In a section entitled "Immunity from, or Liability to, certain Diseases," Hunt quotes extensively from Russell's reports and draws several conclusions. First, the black soldier was mainly different from the white in his greater susceptibility to pneumonia and measles. This was due somewhat to temporary insufficiencies in clothing, food, and shelter but could also be traced to the smaller lung of the black. Prior opinion to the contrary, he showed no immunity to malarial diseases. When in the hospital he was likely to become despondent, to not eat or take medicine unless forced, and to believe magical forces were at work in his case. Hunt believed that this weakness of intellect could be remedied by education and pointed out that free blacks recruited from the north did better as a result.[32]

In the final pages of his paper, Hunt took up the question of the intellectual capacity of the black man. Hunt recognized that "it would be grossly unfair to subject the negro to a comparison of intellectual capacity based on his present manifestation of mental acuteness" since he had been denied education and treated so poorly, especially in the south.[33] And Hunt was not thoroughly convinced that brain size is an adequate reflection of intelligence. Still, he went on to recount the history of such measurement, documenting the inadequacies of each method. None can compare, he argued, with actually weighing the brain at autopsy. "Up to the present war the number of brains carefully weighed by anatomists was small, nor had any attempt been made to educe any difference that might be assigned to race."[34] Then he laid out Russell's autopsy data from 405 autopsies, with a racial census of 24 whites and 381 blacks. Of the black specimens, 241 are designated as being of mixed race, with degree of whiteness indicated. The average weight of the "pure black" brain was forty-seven ounces and the "pure white" was fifty-two ounces. With the exception of the 1/16th white bodies, whose brains weighed forty-five ounces, the remaining mixed-race brains rose in size as the supposed amount of white blood increased.[35]

Although Hunt reported this data, which confirmed contemporary stereotypes of the inferior black mind, he was far from being a pure biological determinist in this paper. He noted that American whites were big-

ger than soldiers in the European countries from which they had come
and that white American brains were heavier than corresponding European ones. "There are evidences," he proclaimed, "that the American, in
founding a new nationality, has also established a new type of manhood."
He was two to three inches taller than the French and English, five to eigh-
teen pounds heavier, and his brain weighed three ounces more. "In fine,
there seems to be some reason to believe that the human brain, in the case
of whites, has been increased in size by its transplantation to this conti-
nent." Bodies and brains could be reshaped by culture, opportunity, and
education. "If it [the white brain] has enlarged under our institutions, why
may not the negro brain, subjected to new and invigorating influences,
also increase its size?"[36] Hunt's message was optimistic; with education,
time, and opportunity, the black man could improve just as the white
American had, although he left unsaid any claims about the potential for
actual equality. Still, it was possible that the black man, too, could achieve
"a new type of manhood."

The final significant publication to describe the soldier's body was Benjamin Gould's *Investigations of the Military and Anthropological Statistics of
American Soldiers*, which appeared in 1869.[37] Gould and others had been
hired by the U.S. Sanitary Commission to measure the soldier and create
tables of normal body measurements. Following Adolphe Quetelet, Gould
and his colleagues sought to describe the average American male in far
greater detail than heretofore imagined. Gould measured parts and the
wholes of arms, legs, feet, heads, and trunks; he gathered heights and
weights; he quantified lifting ability, pulmonary capacity, dental condition, respiration, pulse, and vision. He broke down much of this data by
ethnicity and country of origin, age, and degree of education. Gould measured some 2,020 "full black" and 863 "mixed race" soldiers, and found
a few physical differences. Blacks had smaller pulmonary capacity than
whites, as measured by chest circumference and a spirometer device in-
vented in England two decades before. The black foot was a bit different
in its dimensions, supporting the notion of the flat foot being more common among blacks. And the black soldier had longer arms than the white,
so that the distance from fingertip to kneecap was shorter.

One historian has claimed that in so highlighting this measurement

Gould was marking the black man "as that much closer to anthropoid in development," but Gould himself did not drawn that conclusion.[38] He just documented the measurement. He did note that the mulatto measurements tended not to fall in between black and white but lay outside of the progression, a fact that he believed "cannot fail to attract attention. The well-known phenomenon of their inferior vitality may stand, possibly, in some connection with the fact thus brought to light."[39] Nowhere does Gould comment on his sample, on how these 2,020 men might or might not be representative of the black population in the United States. From Sanitary Commission papers it is clear that at least some of Gould's black soldiers were in New Orleans in 1864 and 1865. Given the distribution of troops late in the war, it is likely that some if not most of these men were ex-slaves. Compared to the Baxter sample, drawn from northern recruiting stations, it is possible that the differences between the Gould and Baxter men in terms of chest circumference and pulmonary function was an accurate measure of prewar health, to the free northern black's advantage. But this is, at best, speculation.

Long after the war was over, these various measurements of the black body continued to influence the intellectual discourse on race and human development. Charles Darwin cited Gould's data in his book *The Descent of Man*, to emphasize the ways in which physical differences among races could be clearly delineated.[40] In her essay on spirometry and race in the nineteenth century, historian Lundy Braun details numerous late nineteenth-century discussions of pulmonary weakness among blacks, culminating in Frederick Hoffmann's analysis of the physical degeneracy of the black race and its coming extinction. Throughout this discourse the measurements of Gould are recounted, while those of Baxter are ignored, in order to establish the physical basis of black inferiority and weakness.[41]

At least as influential was the smaller brain weight recorded by Russell and reported by Hunt. It supported the continuing attempts to rank races by brain size and locate the African on the bottom of the evolutionary order. If one doubted the skull capacity data generated before the war, here was new evidence based on a large number of measurements. Until IQ testing brought a whole new set of numbers into the debate, brain size remained critical in discussions of comparative intelligence.[42] Historian John Haller has argued persuasively that with the results of Civil War anthropometry, "no longer would attitudes of racial inferiority have to employ those prewar measurements and conclusions that had been tainted with proslavery arguments." Instead, conclusions about racial deficits could depend on measurements proven by Civil War data. Haller notes the irony—the war that liberated the American slave also created a body of research used to support institutional racism in the ensuing decades.[43]

Epilogue

Around 180,000 black men went to war; something like 143,000 came home. More than 8,000 of those men received honorable discharge for disabilities, with the most important diagnoses being tuberculosis, chronic diarrhea, rheumatism, and hernias. Some 1,200 men were disabled by gunshot wounds or amputations. Many more went home weakened by infectious diseases and the sequelae of prolonged malnutrition and exposure. These factors no doubt had a negative impact on veterans' postwar earnings. The war may have helped the men, however, exposing them to educational opportunities and broader geographic knowledge. For those men who lived to 1890, the pension program for Union Civil War veterans eased their retirement years and prolonged survival, although white soldiers fared better than their black comrades after the war, just as they had during it.

The discharge of the colored units was often delayed to the end of 1865 and into 1866. When free of the service, many black men returned south, from whence they came, seeking family members and a new start. Not surprisingly, their presence rankled southern whites, especially those men who had fought as Rebels in the war. In Kentucky slavery persisted in the summer of 1865; not until the Thirteenth Amendment went into effect later that year was slavery finally completed banned from the United States. The world these soldiers entered was tumultuous. Southerners needed black labor and sought to replace slavery with a near-bondage equivalent, while ex-slaves sought independence from white hegemony as well as opportunities for political voice, independent livelihoods, and protection from white backlash. The Freedman's Bureau promised support, including hospitals for the sick, but by 1870 it had dissolved, never fulfilling the hopes that the African American community had for aid from it. The black men who had served as soldiers often carried physical disabilities because of the war but also sometimes accumulated pay that helped them buy land or purchase the accoutrements of a trade.[1]

Economic historian Chulhee Lee has examined whether wealth accumulation among white Union veterans was helped or hindered by their war experience. Lee studied a database of Union veterans whose occupational status in 1860 and in 1870 was available from census data. After controlling for various characteristics, he found that "combat exposure and wounds while in service had strong negative effects on the total 1870 wealth." Illness during the war, on the other hand, had no significant effect. He further found that the impact of wealth loss caused by wounds was most severe for veterans who were unskilled workers at the time of enlistment. The effect of leg and foot injuries was particularly strong, as one might expect. It is hard to dig ditches, load boxes, or work construction when unable to bear weight.[2] In a sample of black veterans who survived long enough to enter the pension system, my colleagues and I found that 93 percent of the men were working at some form of manual labor, including farming and unskilled or semiskilled jobs.[3] It is likely that the disabled black veteran was unable to find gainful employment, especially if damaged in a lower extremity. But he did have the additional income from a pension if he was savvy enough to know how to apply for one.

The war may have expanded the black man's range of choices in occupation and location. For the 140,000 recruited directly from slave plantations or contraband camps, this statement is almost laughable, as it contrasts slavery with freedom. But beyond this massive transition (true for the soldier as well as the nonsoldier), particular benefits accrued to some African American soldiers. Conscientious commanders demanded that education be part of the soldier's day, with particular emphasis on reading and rudimentary arithmetic. It is hard to know how many soldiers benefitted from this exposure, but everyone who learned to read was that much more able on the job market when he left the army. Another factor in army life which offered the soldier an advantage was his exposure to different parts of the country and contacts in those disparate areas. The slave who had known only his small region in Alabama, for example, commingled with black men from New York and Philadelphia and learned about opportunities in those cities. Economic historians Dora Costa and Matthew Kahn have found that men who were in mixed regiments, regiments that were diverse as to geographic origin, were more likely to move across states and across census regions after the war. Soldiers who traveled to a region new to them during the war were also more likely to move from the area of their nativity. If they had never seen a city before the war and traveled there while in uniform, they were more likely to move to a city after 1865. Wartime exposures opened a world of possibilities to the black soldier, even it remained hedged in by race and class.[4]

Some black men used their army experience to forward their careers. Alexander Augusta had been prominent during the war in arguing for the rights of black men, particularly black physicians. After the war he participated in the founding of Howard Medical School and led the fight for admission of black physicians to the American Medical Association.[5] In Virginia, J. Dennis Harris parlayed his service during the war as a physician to black troops and freedmen into a political career. He ran for lieutenant governor in 1869 and lost only by a slim margin (in an election with open black franchise and from which many white former Confederates were excluded).[6] Several black commissioned officers of the 55th Massachusetts received federal patronage jobs in Boston, and other men built on prewar connections with prominent black abolitionists like Frederick Douglass and Martin Delany. Lewis Latimer worked in a patent-attorney office after returning from his stint in the U.S. navy. There he acquired mechanical drawing skills that led to his employment with Alexander Graham Bell and Thomas Edison.[7]

It is difficult to say whether ex-soldiers fared better, worse, or the same as able-bodied black men who left slavery at the end of the war without any army service. Finding a comparison group with recorded data analogous to that available for the soldiers is difficult. It is clear, however, that black soldiers continued to have higher mortality rates after the war than white veterans. According to Costa, black veterans "faced an odds of dying at older ages that was 1.3 times higher than that of white Union Army veterans and black men did not achieve the older age mortality rates of white Union Army veterans until the 1970s. Deaths from infectious disease accounted for half of the black-white mortality gap."[8] Costa further found that black veterans living in urban areas, those very men most likely

FRANCHISE,
AND NOT THIS MAN?"

This sympathetic illustration of a wounded black soldier emphasizes the price
he has paid for the right to citizenship and the vote. *Harper's Weekly*, 5 August
1865.

to benefit from the economic opportunity of northern cities, also faced a
15 percent mortality deficit as a result. By 1900 a little over 60 percent of
the white veterans who had survived the war were still alive, whereas only
40 percent of the black veterans remained. This may reflect persistent sequelae of the greater disease burden black soldiers carried during the war.
It is also very likely that it is a reflection of the class differences among

black and white soldiers, with all the associated disease differential brought by poverty, malnutrition, and overcrowding.

It is equally hard to assess what difference black soldiers made to the war effort. One can debate for hours (or pages) about why the South lost the war or why the North won. Many historians have spilled ink on this issue, and new books addressing it continue to appear. The most common factors listed include the northern superiority in agricultural and industrial production, the inability of the South to win foreign recognition and support due to emerging European repugnance about slavery, and the failure of Jefferson Davis's government to elicit a truly national effort from states committed to states rights.[9] One novelist, Harry Turtledove, has the war hinge on the Confederate battle orders that fell into Union hands before Antietam. He changes that moment's action, the South wins at Antietam, Lincoln does not issue the Emancipation Proclamation, Europe comes in on the side of the South, Lincoln loses the fall 1864 election, and the Confederacy sues successfully for peace from the subsequent Copperhead administration. In Turtledove's counterfactual world, the British government requires the Confederacy to commit to liberating the slaves as a condition of their support, and by the end of the war the Confederacy has actually enlisted slaves into their armies, with promises of freedom and other postwar benefits.[10]

The enlistment of black troops affected these factors in the war's outcome in a variety of ways. The contraband policy and subsequent enrollment of ex-slaves into the army may well have encouraged slaves to escape, knowing they would be welcomed across Union lines. Abraham Lincoln may not have been entirely correct when he told Andrew Johnson, "The bare sight of 50,000 armed and drilled black soldiers upon the banks of the Mississippi, would end the rebellion at once," but the presence of black troops must have lowered southern morale, especially among those who harbored beliefs about the "happy darkies."[11] With three-quarters of the black soldiers coming directly from southern plantations, the black enlistment effort simultaneously drained the southern agricultural labor force while increasing the northern armed strength. Robert E. Lee's army dwindled from desertion in the last months of the war; many men went home because their families were starving and the fields were barren. There just were not enough men left in the South to plant crops and fight battles, too. The enlistment of black soldiers contributed significantly to that outcome. Even the Confederate government came to realize that freeing and arming black slaves was the only way that the army could be rejuvenated, but this decision came too late to prevent the surrender at Appomattox.[12]

The war might have been over sooner had the Union taken better care of the black troops, but it is clear than many of the men in charge never regarded their black soldiers as worthy of full respect. As a result, black units mostly functioned below their full potential as so many men were downed or weakened by disease. Had they received even the minimal care meted out to white troops, the same (though still inadequate) diet, the same amount of fatigue duty, the same uniforms, the same tents, and had they been led by experienced, caring officers, much of the disease that mowed them down could have been prevented. The army as a whole would have benefited. Racism, like the oppression of women, results in the wastage of human potential, and nowhere was this more true than in the regiments of the U.S. Colored Troops.

It took the U.S. Army a long time to learn this lesson. The *New York Times* admitted after the war was over that the value of black troops had been proven beyond a doubt. "It is evident from our experience that . . . the negroes are capable of being transformed into soldiers; that they can fight, and can do efficient work in all arms of service." They had proven their loyalty and devotion to country. "We have tried the experiment of African soldiers, in a civilized army, upon a scale of magnitude never before attempted or approached in the history of the world."[13] Not until the Korean War would the U.S. Army realize that segregated troops would always be separate and unequal and finally integrate its forces. In the United States today the army stands as a model of successful integration, with black men serving all the way to the top of the ranks.

It took medical science even longer to recognize that concepts like racial degeneracy or inherent lack of endurance had no place in scientific thinking. Discussing racial weakness as an answer to the reasons for high mortality among black troops allowed the army's medical department to ignore remediable causes and miss valuable lessons from the grim statistics. Only in the past two or three decades have racial disparities in health care occupied a significant place in medical research in the United

States, now with an emphasis instead on discovering the social, economic, and environmental causes that create different outcomes among ethnic groups. The current pendulum may have swung too far toward those who deny any significant biological or genetic differences among ethnic groups, but it is a corrective that has been long needed.

In the movie *Glory*, which tells the story of the 54th Massachusetts Infantry Regiment (colored), the character of Robert Gould Shaw struggles to create a regiment of disciplined men who acknowledge his authority. After Colonel Shaw rebukes a man for being too slow to load his rifle, Maj. Cabot Forbes asks why Shaw is so harsh toward his fledgling soldiers.

> Major Forbes: Why do you treat the men this way Robert?
> Colonel Robert G. Shaw: How should I treat them?
> Major Forbes: . . . Like men?[14]

The remainder of the movie reminds us that whatever the white officer thought, the African American soldier did not require white validation to prove his manhood.

Abbreviations

Berlin, *Freedom:* Ira Berlin, ed., Joseph P. Reidy and Leslie S. Rowland, assoc. eds. *Freedom: A Documentary History of Emancipation, 1861–1867, Selected from the Holdings of the National Archives of the United States. Series II: The Black Military Experience.* Cambridge: Cambridge University Press, 1982. This 852-page volume reproduces documents from the National Archives, with editorial commentary.

MSHW: *Medical and Surgical History of the War of the Rebellion (1861–1865).* 12 vols. and 3 index vols. Washington: Government Printing Office, 1870–83; facsimile rpt., Wilmington, N.C.: Broadfoot Publishing Co., 1990.

NARA: National Archives and Records Administration. All NARA documents reviewed for this book are located at the National Archives building on the Mall in Washington, D.C.

NARA M858: A five-reel document collection gathered by archivists in the National Archives in the 1880s and later microfilmed as "The Negro in the Military Service of the United States, 1639–1886." NARA Microfilm Publication M858, Record Group 94. National Archives, Washington, D.C.

OR: *The War of the Rebellion: A Compilation of the Official Records of the Union and Confederate Armies.* Washington: Government Printing Office. Ser. 1, 1–53; Ser. 2, 1–8; Ser. 3, 1–5; Ser. 4, 1–4 (1880–1901). Now available and searchable on-line at *http://cdl.library.cornell.edu/moa/browse.monographs/waro.html* (accessed August 29, 2006). These volumes consist of various official documents saved from the war and reprinted together in the 1880s and 1890s.

USSC Papers: Microfilmed collection of the U.S. Sanitary Commission Records, Ser. 1, Medical Committee Archives, 1861–1865. Wilmington, Del.: Scholarly Resources Inc., 1998. The originals are held in the Manuscripts and Archives Division, The New York Public Library, Astor, Lenox and Tilden Foundations, New York, New York.

Preface

1. Robert Benchley, "How to Get Things Done," in *The Benchley Roundup*, ed. Nathaniel Benchley (Chicago: University of Chicago Press, 1954), 5–10.

2. "Ira Russell," in *Physicians and Surgeons of the United States*, ed. William B. Atkinson (Philadelphia: C. Robson, 1878), 310–311; Ira Russell to Henry Wilson, 20 December 1863, frames 778–780, reel 10, USSC Papers.

3. USSC Papers.

4. Joseph T. Glatthaar, "The Costliness of Discrimination: Medical Care for Black Troops in the Civil War," in *Inside the Confederate Nation: Essays in Honor of Emory M. Thomas*, ed. Lesley J. Gordon and John C. Inscoe (Baton Rouge: Louisiana State University Press, 2005), 251–271; Paul E. Steiner, *Medical History of a Civil War Regiment: Disease in the Sixty-Fifth United States Colored Infantry* (Clayton, Mo.: Institute of Civil War Studies, 1977).

5. See, e.g., Brian D. Smedley, Adrienne Y. Stith, and Alan R. Nelson, eds., *Unequal Treatment: Confronting Racial and Ethnic Disparities in Health Care* (Washington, D.C.: National Academies Press, 2003).

6. For one thoughtful discussion of these labels, see Orlando Patterson, *Rituals of Blood: Consequences of Slavery in Two American Centuries* (Washington, D.C.: Civitas/Counterpoint, 1998).

7. *MSHW*, vol. 1.

8. The memorial's web site is www.afroamcivilwar.org (accessed 21 March 2007).

9. *MSHW*.

10. Steiner, *Medical History*.

11. Donald Shaffer discusses the fluidity of black names in *After the Glory: The Struggles of Black Civil War Veterans* (Lawrence: University Press of Kansas, 2004).

12. Charles Rosenberg, "Framing Disease," in *Framing Disease: Studies in Cultural History*, ed. Charles Rosenberg and Janet Golden (New Brunswick, N.J.: Rutgers University Press, 1987), xiii–xxvi.

13. See chapter 7.

14. Thomas Wentworth Higginson, *Army Life in a Black Regiment* (1870; rpt., Lansing: Michigan State University Press, 1960), 190.

One · The Black Body at War

1. Robert Durden, *The Gray and the Black: The Confederate Debate on Emancipation* (Baton Rouge: Louisiana University Press, 1972).

2. Joseph T. Glatthaar, *Forged in Battle: The Civil War Alliance of Black Soldiers and White Officers* (Baton Rouge: Louisiana State University Press, 1990); Keith Wilson, *Campfires of Freedom: The Camp Life of Black Soldiers during the Civil War* (Kent, Ohio: Kent State University Press, 2002); John David Smith, ed., *Black Soldiers in Blue: African American Troops in the Civil War Era* (Chapel Hill: University of North Carolina Press, 2002); Noah Andre Trudeau, *Like Men of War: Black Troops in the Civil War, 1862–1865* (Edison, N.J.: Castle Books, 2002); Benjamin Quarles, *The Negro in the Civil War* (Boston: Little, Brown, 1953); Howard C. Westwood, *Black Troops, White Commanders, and Freedmen during the Civil War* (Carbondale, Ill.: Southern Illinois University Press, 1992); James M. McPherson, *The Negro's Civil War: How American Blacks Felt and Acted during the War for the Union* (New York: Vintage, 1965); Ira Berlin, Joseph P. Reidy, and Leslie S. Rowland, eds., *Freedom's Soldiers: The Black Military Experience in the Civil War* (Cambridge: Cambridge University Press, 1998); Russell Duncan, *Where Death and Glory Meet: Robert Gould Shaw and the 54th Massachusetts Infantry* (Athens: University of Georgia Press, 1999).

3. Abraham Lincoln to Horace Greeley, 22 August 1862, reproduced in frame 602, reel 1, NARA M858.

4. James M. McPherson, *Battle Cry of Freedom: The Civil War Era* (New York: Ballantine, 1988), 490–510, 545–559.

5. McPherson, *Battle Cry*, 355–356.

6. Glatthaar, *Forged in Battle*, 4–9.

7. Samuel L. Kirkwood to Henry Halleck, 5 August 1862, reproduced in frame 933, reel 1, NARA M858.

8. Order issued 17 July 1862, by order of the Secretary of War, under the auspices of E. D. Townsend, Assistant Adjutant General, frames 914–915, reel 1, NARA M858.

9. Ibid.

10. E. D. Townsend, 3 October 1863, General Order 329, gave details of recruitment of current slaves; reproduced in frames 1654–1656, reel 3, NARA M858. W. Bradford, Governor of Maryland, to Montgomery Blair, 11 September 1863 (reproduced in frames 1569–1570, reel 2, NARA M858), protested recruitment of soldiers off plantations and quartering of black soldiers in neighborhood.

11. James M. Fidler to W. H. Sidell, 15 June 1865, described conditions of recruitment in Kentucky; reproduced in frames 3666–3672, reel 4, NARA M858. General Orders 33, 11 March 1865, printed the congressional resolution that freed families of enlisted slaves, in frame 3595, reel 4, NARA M858.

12. Trudeau, *Like Men of War,* map on inside of front cover.

13. Charles A. Dana to Edwin M. Stanton, 22 June 1863, reproduced in frame 1343, reel 2, NARA M858.

14. W. T. Sherman to Lorenzo Thomas, telegram 21 June 1864, reproduced in frame 2639, reel 3, NARA M858.

15. Glatthaar, *Forged in Battle;* Wilson, *Campfires of Freedom.*

16. J. H. Baxter, comp., *Statistics, Medical and Anthropological of the Provost-Marshal-General's Bureau, Derived from Records of the Examination for Military Service during the Late War of the Rebellion of over a Million Recruits, Drafted Men, Substitutes, and Enrolled Men* (Washington: Government Printing Office, 1875), 31–32.

17. Robert William Fogel and Stanley L. Engerman, *Time on the Cross: The Economics of American Negro Slavery,* 2 vols. (Boston: Little, Brown, 1974).

18. See, e.g., Richard Sutch, "The Treatment Received by American Slaves: A Critical Review of the Evidence Presented in 'Time on the Cross,'" *Explorations in Economic History* 12 (1975): 335–348; and Kenneth Kiple and Virginia Kiple, "Slave Child Mortality: Some Nutritional Answers to a Perennial Puzzle," *Journal of Social History* 10 (1977): 284–309.

19. Richard H. Steckel, "A Peculiar Population: The Nutrition, Health, and Mortality of American Slaves from Childhood to Maturity," *Journal of Economic History* 46 (1986): 721–741. Todd L. Savitt describes the many diseases that could have led to these poor growth measurements in *Medicine and Slavery: The Diseases and Health Care of Blacks in Antebellum Virginia* (Urbana: University of Illinois Press, 1978).

20. Dora L. Costa, "The Measure of Man and Older Age Mortality: Evidence from the Gould Sample," *Journal of Economic History* 64 (2004): 1–23.

21. Chulhee Lee, "Prior Exposure to Disease and Later Health and Mortality: Evidence from the Civil War Medical Records," in *Health and Labor Force Participation over the Life Cycle: Evidence from the Past,* ed. Dora L. Costa (Chicago: University of Chicago Press, 2003); and Chulhee Lee, "Socioeconomic Background, Disease, and Mortality among Union Army Recruits: Implications for Economic and Demographic History," *Explorations in Economic History* 34 (1997): 27–55.

22. Sven Wilson and Clayne L. Pope, "The Height of Union Army Recruits: Family and Community Influences," in Costa, *Health and Labor Force Participation,* 113–145.

23. Baxter, *Statistics,* 22.

24. Ibid.

25. Ira Russell, "Report on Hospital L'Ouverture," Alexandria, Va., frame 285, reel 3, USSC Papers.

26. *MSHW,* vol. 1, xxxvi–xxxix. There are many sources of imprecision in these numbers, which are detailed in this section of the *MSHW.*

27. Paul E. Steiner, *Medical History of a Civil War Regiment: Disease in the Sixty-*

Fifth United States Colored Infantry (Clayton, Mo.: Institute of Civil War Studies, 1977), xvii.

28. Steiner, *Medical History*, xvii, 43–50.

Two • Pride of True Manhood

1. Anxieties about the liberated slaves and their behavior is evident in the hearings and reports of the commission. See American Freedmen's Inquiry Commission, *Preliminary Report Touching the Condition and Management of Emancipated Refugees; Made to the Secretary of War, June 30, 1863* (New York: John F. Trow, 1863); Reports and testimony before the American Freedmen's Inquiry Commission, nos. 199, 200, and 201, microfilm set M619, NARA. On attitudes toward miscegenation, see Joel Williamson, *New People: Miscegenation and Mulattoes in the United States* (New York: Free Press, 1980), and Elise Lemire, *"Miscegenation": Making Race in America* (Philadelphia: University of Pennsylvania Press, 2002). Mary Chestnut was particularly disgusted with her father-in-law and all of his light-brown offspring; see Mary Chestnut, *Mary Chestnut's Civil War*, ed. C. Vann Woodward (New Haven: Yale University Press, 1981).

2. Joseph R. Smith, "Sanitary Report of the Dept of Arkansas for the Year 1864," manuscript number MS C 126, Historical Collection, National Library of Medicine, Bethesda, Md.

3. Todd Balf, *The Darkest Jungle: The True Story of the Darién Expedition and America's Ill-Fated Race to Connect the Seas* (New York: Crown, 2003).

4. J. T. Headley, "Darien Exploring Expedition, under the Command of Lieut. Isaac C. Strain," *Harper's New Monthly Magazine* 10 (1855): 433–458, 600–615, 745–764.

5. Ibid., 458.

6. Ibid., 761.

7. Smith, "Sanitary Report."

8. Headley, "Darien Exploring Expedition," 451–452.

9. Stephen W. Berry II, *All That Makes a Man: Love and Ambition in the Civil War South* (New York: Oxford University Press, 2003).

10. The USSC questionnaire is in frame 113–114, reel 1, USSC Papers.

11. F. E. Pequette to O. M. Long, 13 November 1865, frame 568, reel 5, USSC Papers.

12. Wm. M. Brown to Ira Russell, 10 October 1865, frame 1137, reel 5, USSC Papers.

13. B[enjamin] Woodward, "Report on the Diseases of Colored Troops," frame 897, reel 2, USSC Papers.

14. H. B. Hubbard, comments in *Statistics, Medical and Anthropological of the Provost-Marshal-General's Bureau, Derived from Records of the Examination for Mili-*

tary Service during the Late War of the Rebellion of Over a Million Recruits, Drafted Men, Substitutes, and Enrolled Men, comp. J. H. Baxter (Washington: Government Printing Office, 1875), 199.

15. C. L. Hubbell, comments in ibid., 261.

16. Robert Carswell, comments in ibid., 183

17. C. W. Buckley to L. Thomas, 1 April 1865, in Berlin, *Freedom,* 564.

18. Martin Robison Delany, *The Condition, Elevation, Emigration, and Destiny of the Coloured People of the United States* (1852; rpt., New York: Arno, 1968), 50 and 86.

19. David Walker, *Appeal to the Colored Citizens of the World,* 1829, as quoted in Mia Bay, *The White Image in the Black Mind: African American Ideas about White People, 1830–1925* (New York: Oxford University Press, 2000), 41. The quote from Bay herself is also on p. 41.

20. Delany, *The Condition,* 10.

21. On the black intellectual's project of self-elevation, see Patrick Rael, *Black Identity and Black Protest in the Antebellum North* (Chapel Hill: University of North Carolina Press, 2002); Eddie S. Glaude, *Exodus! Religion, Race and Nation in Early Nineteenth-Century Black America* (Chicago: University of Chicago Press, 2000); and Wilson Jeremiah Moses, *The Golden Age of Black Nationalism, 1850–1925* (Hamden, Conn.: Archon, 1978).

22. Many sources describe the grim conditions of chattel slavery in the United States. See David Brion Davis, *Inhuman Bondage: The Rise and Fall of Slavery in the New World* (Oxford: Oxford University Press, 2006). It not only is a magisterial synthesis but also contains references to the major works on this central topic in American history.

23. Bay, *The White Image,* 127–143.

24. George M. Frederickson, *The Black Image in the White Mind: The Debate on Afro-American Character and Destiny, 1817–1914* (New York: Harper and Row, 1971), 43–70.

25. Ibid., 71–96; John S. Haller, *Outcasts from Evolution: Scientific Attitudes of Racial Inferiority, 1859–1900* (Urbana: University of Illinois Press, 1971); William Stanton, *The Leopard's Spots: Scientific Attitudes toward Race in America, 1815–59* (Chicago: University of Chicago Press, 1960).

26. Deborah Gray White, *Ar'n't I a Woman? Female Slaves in the Plantation South,* rev. ed. (New York: Norton, 1999).

27. See, e.g., "Another Horrid Massacre by Negro Soldiers," *Missouri Republican,* 6 September 1863. In spite of its name, this newspaper opposed the recruitment of black soldiers and frequently expressed racist ideology.

28. Alexander Crummell, introduction to Edward W. Blyden, *A Vindication of the African Race; Being a Brief Examination of the Arguments in Favor of African Inferiority* (Monrovia, Liberia: Gaston Killian, 1857), 5.

29. Lemire, *"Miscegenation,"* 87–114.

30. Joel Williamson, "The Soul is Fled," in *New Perspectives on Race and Slavery in America: Essays in Honor of Kenneth M. Stampp*, ed. Robert H. Abzug and Stephen E. Maizlish (Lexington: University Press of Kentucky, 1986), 185–198, quote on p. 186.

31. Cited in Berry, *All that Makes a Man*, 139.

32. Rowland Berthoff, "Conventional Mentality: Free Blacks, Women, and Business Corporations as Unequal Persons, 1820–1870," *Journal of American History* 76 (1989): 753–784.

33. Keith Wilson, *Campfires of Freedom: The Camp Life of Black Soldiers during the Civil War* (Kent, Ohio: Kent State University Press, 2002).

34. See Frederickson, *The Black Image*, 5, 12–21.

35. Joshua D. Rothman, *Notorious in the Neighborhood: Sex and Families across the Color Line in Virginia, 1787–1861* (Chapel Hill: University of North Carolina Press, 2003), 93–129.

36. Berthoff, "Conventional Mentality," 771.

37. Frederick Douglass, "The Claims of the Negro Ethnologically Considered," 1854, quoted in Bay, *The White Image*, 69.

38. Frederick Douglass, "Enlistment of Colored Men," *Douglass Monthly*, August 1863, 850–852, quote on p. 852.

39. Ibid.

40. Unknown officer of the 22nd U. S. Colored Infantry, letter to the *Philadelphia Press*, reprinted in the *Liberator*, 22 July 1864, reprinted in James M. McPherson, *The Negro's Civil War: How American Blacks Felt and Acted during the War for the Union* (1965; rpt., New York: Vintage, 1991), 228.

41. Quoted in Joseph Glatthaar, *Forged in Battle: The Civil War Alliance of Black Soldiers and White Officers* (New York: Free Press, 1990), 79.

42. "African Encroachments," *Missouri Republican*, 21 November 1863.

43. The letter from Stanton appointing Robert Dale Owen, James McKay, and Samuel Gridley Howe to the commission is in the *OR*, ser. 3, vol. 3, p. 73. See also American Freedmen's Inquiry Commission, *Preliminary Report*, and Reports and testimony before the American Freedmen's Inquiry Commission, reels 199, 200, and 201, NARA microfilm M619.

44. "Final Report of the American Freedmen's Inquiry Commission to the Secretary of War," 15 May 1864, in the *OR*, ser. 3, vol. 4, 289–382. On the Freedmen's Bureau hospitals in one southern state, see Todd L. Savitt, "Politics in Medicine: The Georgia Freedmen's Bureau and the Organization of Health Care, 1865–1866," *Civil War History* 28 (1982): 45–64.

45. Wilson, *Campfires of Freedom*, xv.

46. General Orders No. 13, Head Quarters 49th USCI, 11 Apr 1865, Vicksburg, Miss., in Berlin, *Freedom*, 624–625.

47. Wilson, *Campfires of Freedom*, esp. 58–59.

48. Quoted in Berry, *All That Makes a Man*, 176–177.

49. Sam R. Watkins, *"Co. Aytch": Maury Grays First Tennessee Regiment or A Side Show of the Big Show* (1882; rpt., Wilmington, N.C.: Broadfoot Publishing Co., 1994), 71–72.

50. Joseph T. Glatthaar, "'Glory,' the 54th Massachusetts Infantry, and Black Soldiers in the Civil War," *History Teacher* 24 (1991): 475–485.

51. Jonathan Walters, "Invading the Roman Body: Manliness and Impenetrability in Roman Thought," in *Roman Sexualities*, ed. Judith P. Hallet and Marilyn B. Skinner (Princeton: Princeton University Press, 1997), 29–43. My thanks to Lynda Coon, historian of the Dark Ages at the University of Arkansas, who called to my attention this reference and the cultural meanings of the lash.

52. Thomas Wentworth Higginson, *Army Life in a Black Regiment* (1870; rpt., Lansing: Michigan State University Press, 1960), 22 and 24.

53. Jules Zanger, "The 'Tragic Octoroon' in Pre-Civil War Fiction," *American Quarterly* 18 (1966): 63–70; Stephan Talty, *Mulatto America: At the Crossroads of Black and White Culture: A Social History* (New York: HarperCollins Books, 2003); James Kinney, *Amalgamation! Race, Sex and Rhetoric in the Nineteenth-Century American Novel* (Westport, Conn.: Greenwood, 1985); Eve Allegra Raimon, *The "Tragic Mulatta" Revisited: Race and Nationalism in Nineteenth-Century Antislavery Fiction* (New Brunswick, N.J.: Rutgers University Press, 2004). A white planter attempts to rape his own mulatta daughter in the best-selling novel, *The White Slave: A Story of Life in Virginia*, by R. Hildreth, which appeared in 1852. Gunther Peck dissects the white slave metaphor in "White Slavery and Whiteness: A Transnational View of the Sources of Working-Class Radicalism and Racism," *Labor Studies in Working-Class History of the Americas* 1 (2004): 41–63.

54. Rothman, *Notorious in the Neighborhood*.

55. Department of Commerce, Bureau of the Census, *Negro Population, 1790–1915* (Washington: Government Printing Office, 1918), 220–221; Williamson, *New People*.

56. Charles Sumner, *The Barbarism of Slavery*, 1860, quoted in Robert Brent Toplin, "Between Black and White: Attitudes toward Southern Mulattoes, 1830–1861," *Journal of Southern History* 45 (1979): 185–200, quote on p. 189.

57. Williamson, *New People*, 72; Josiah Clark Nott, "Statistics of Southern Slave Population, with Especial Reference to Life Insurance," *De Bow's Review* 4 (1847): 273–287.

58. Sanitary Commission survey, frames 113–114, reel 1, USSC Papers.

59. Department of Commerce, *Negro Population*, 220–221; the autopsy records are in the USSC Papers, frames 200–269, reel 5. The autopsy reports consist of 452 names, divided into two subsets. The first 200 are in one handwriting; the second 252 in another. Skin color was recorded by fraction of white blood (e.g., 1/8 white). In the first set there were only eight men with 0 percent white blood. In the second cluster, on the other hand, blk or bk or B appears in 98 of 252 cases.

Does this represent an actual change in the ethnic mix of the men or a change in how their skin tone was assessed? The answer is unknown.

60. Ira Russell, Hygienic and Medical Notes, frame 282, reel 5, USSC Papers.

61. Robert A. Margo and Richard H. Steckel, "The Heights of American Slaves: New Evidence on Slave Nutrition and Health," *Social Science History* 6 (1982): 516–538.

62. Ira Russell, "Report on Hospital L'Ouverture," Alexandria, Va., frames 290–291, reel 5, USSC Papers.

63. Ibid., frames 291–292, 307.

Three · Biology and Destiny

1. Todd Savitt describes these assumptions in *Medicine and Slavery: The Diseases and Health Care of Blacks in Antebellum Virginia* (Urbana: University of Illinois Press, 1978), especially in chapter 1. Historian John S. Haller Jr. has thoroughly documented and explored these ideas of physical difference (and, by and large, inferiority) in several of his writings. See his *Outcasts from Evolution: Scientific Attitudes of Racial Inferiority, 1859–1900* (New York: McGraw-Hill, 1971) and the articles "Civil War Anthropometry: The Making of a Racial Ideology," *Civil War History* 16 (1970): 309–324; "The Negro and the Southern Physician: A Study of Medical and Racial Attitudes 1800–1860," *Medical History* 16 (1972): 238–253; "Concepts of Race Inferiority in Nineteenth-Century Anthropology," *Journal of the History of Medicine and Allied Sciences* 25 (1970): 40–51.

2. Abraham Lincoln, Fourth Lincoln-Douglas Debates, 15 September 1858, available at www.classic-literature.co.uk/american-authors/19th-century/abraham-lincoln/the-writings-of-abraham-lincoln-04/ (accessed 26 July 2006).

3. Position paper, American Anthropology Association, published in *American Anthropologist* 100 (1998): 712–713.

4. David Brion Davis, "Constructing Race: A Reflection," *William and Mary Quarterly* 54 (1997): 7–18.

5. See, e.g., Joseph L. Graves, *The Emperor's New Clothes: Biological Theories of Race at the Millennium* (New Brunswick, N.J.: Rutgers University Press, 2001), 1.

6. Stephen J. Gould, *The Mismeasure of Man* (New York: W. W. Norton, 1981).

7. Richard J. Herrnstein and Charles Murray, *The Bell Curve: Intelligence and Class Structure in American Life* (New York: Free Press, 1994); Russell Jacoby and Naomi Glauberman, *The Bell Curve Debate: History, Documents, Opinions* (New York: Times Books, 1995).

8. Barbara J. Fields, "*Origins of the New South* and the Negro Question," *Journal of Southern History* 67 (2001): 811–826, quote on p. 816.

9. NIH Policy on Reporting Race and Ethnicity Data: Subjects in Clinical Research, released 8 August 2001, available at *http://grants.nih.gov/grants/guide/notice-files/NOT-OD-01–053.html* (accessed 30 March 2007).

10. Editorial, *News and Observer*, 1 December 2002.

11. Anonymous, "Summary," in Ian D. Smedley, Adrienne Y. Stith, and Alan R. Nelson, eds., *Unequal Treatment: Confronting Racial and Ethnic Disparities in Health Care* (Washington, D.C.: National Academies Press, 2003), 1.

12. Ibid., 8.

13. Robert S. Schwartz, "Racial Profiling in Medical Research," *New England Journal of Medicine* 344 (2001): 1392–1393.

14. Luca Cavalli-Sforza, Pablo Menozzi, and Alberto Piazza, *The History and Geography of Human Genes* (Princeton: Princeton University Press, 1994).

15. Neil Risch, E. Burchard, E. Ziv, and H. Tang, "Categorization of Humans in Biomedical Research: Genes, Race and Disease," *Genome Biology* 3 (2002): 2007.1–2007.12, quote on 2007.1.

16. See Robert Blakely and Judith M. Harrington, eds., *Bones in the Basement: Postmortem Racism in Nineteenth Century Medical Training* (Washington, D.C.: Smithsonian Institution Press, 1997).

17. See, e.g., H. Jack Geiger, "Racial and Ethnic Disparities in Diagnosis and Treatment: A Review of the Evidence and a Consideration of Causes," in Smedley, Stith, and Nelson, *Unequal Treatment*, 417–454.

18. These ideas are widespread in the biological, medical, and historical literature. See, e.g., Paul Ewald, *Evolution of Infectious Disease* (New York: Oxford University Press, 1994); Jared Diamond, *Guns, Germs, and Steel: The Fate of Human Societies* (New York: W. W. Norton, 1997); William H. McNeill, *Plagues and Peoples* (New York: Doubleday, 1977); Alfred W. Crosby, *The Columbian Exchange: Biological and Cultural Consequences of 1492* (Westport, Conn.: Greenwood, 1972); and Gerald Grob, *The Deadly Truth: A History of Disease in America* (Cambridge, Mass.: Harvard University Press, 2002).

19. Margaret Humphreys, *Malaria: Poverty, Race and Public Health in the United States* (Baltimore: Johns Hopkins University Press, 2001).

20. Anonymous editorial, "Endurance of Black Soldiers," *Weekly Anglo-African*, 5 September 1863, 1.

21. *MSHW*, 5:80–89; quotation on p. 85. In the Spanish-American War researchers found that most typho-malarial fever was actually typhoid fever, but it is hard to know what this diagnosis meant in the context of the Civil War. See Vincent Cirillo, *Bullets and Bacilli: The Spanish-American War and Military Medicine* (New Brunswick, N.J.: Rutgers University Press, 2004).

22. Ira Russell, "Report on Hospital L'Ouverture," frame 299, reel 3, USSC Papers.

23. Humphreys, *Malaria*.

24. Ibid.

25. Margaret Humphreys, *Yellow Fever and the South* (New Brunswick, N.J.: Rutgers University Press, 1992), 27.

26. Elisha Harris, "Yellow Fever on the Atlantic Coast and at the South during the War," in *Contributions Relating to the Causation and Prevention of Disease and to Camp Diseases; Together with a Report of the Diseases, Etc., among the Prisoners at Andersonville, Ga.*, ed. Austin Flint (New York: Hurd and Houghton, 1867), 236–268; Thomas J. Farmham and Francis P. King, "'The March of the Destroyer': The New Bern Yellow Fever Epidemic of 1864," *North Carolina Historical Review* 73 (1996): 435–483.

27. Charles Higham, *Murdering Mr. Lincoln: A New Detection of the 19th Century's Most Famous Crime* (Beverly Hills, Calif.: New Millennium Press, 2004), p. 58.

28. *MSHW*, 1:636–637, 710.

29. Sheldon Watts attacks Kiple's ideas in "Yellow Fever Immunities in West Africa and the Americas in the Age of Slavery and Beyond: A Reappraisal" and "Response to Kenneth Kiple," *Journal of Social History* 34 (2001): 955–967, 975–976. Kiple has made this argument about innate immunities for more than two decades. He reviews it (and cites his earlier publications) in "Response to Sheldon Watts, 'Yellow Fever Immunities in West Africa and the Americas in the Age of Slavery and Beyond: A Reappraisal,'" in *Journal of Social History* 34 (2001): 969–974.

30. Ira Russell, "Report on Hospital L'Ouverture," frame 286, reel 3, USSC Papers.

31. George Andrew to Elisha Harris, 17 July 1865, frame 413, reel 1, USSC Papers.

32. Benjamin Woodward, "Report on the Diseases of Colored Troops," frame 890, reel 2, USSC Papers.

33. Ira Russell, "Report on Hospital L'Ouverture," frame 284, reel 3, USSC Papers.

34. *MSHW*, 1:637 and 705.

35. Lundy Braun, "Spirometry, Measurement, and Race in the Nineteenth Century," *Journal of the History of Medicine* 60 (2005): 135–169, argues that the social construction of these categories and their measurement makes the entire enterprise suspect. On modern research into racial disparities and lung capacity, see, e.g., C. L. Joseph, D. R. Ownby, E. L. Peterson, and C. C. Johnson, "Racial Differences in Physiologic Parameters Related to Asthma among Middle-Class Children," *Chest* 117 (2000): 1336–1344; J. L. Hankinson, John R. Odencrantz, and Kathy B. Fedan, "Spirometric Reference Values from a Sample of the General U. S. Population," *American Journal of Respiratory and Critical Care Medicine*, 159 (1999): 179–187; P. L. Enright, A. Arnold, T. A. Manolio, and L. H. Kulter, "Spirometry Reference Values for Healthy Elderly Blacks: The Cardiovascular Health Study Research Group," *Chest* 110 (1996): 1416–1424.

36. B. J. Marston et al., "Incidence of Community-Acquired Pneumonia Re-

quiring Hospitalization: Results of a Population-Based Active Surveillance Study in Ohio: The Community-Based Pneumonia Incidence Study Group," *Archives of Internal Medicine* 157 (1997): 1709–1718.

37. M. D. Redelings, F. Sorvillo, and P. Simon, "A Population-Based Analysis of Pneumococcal Disease Mortality in California, 1989–1998," *Public Health Reports* 120 (2005): 157–164; J. S. Haas, M. L. Dean, Y. Hung, D. J. Rennie, "Differences in Mortality among Patients with Community-Acquired Pneumonia in California by Ethnicity and Hospital Characteristics," *American Journal of Medicine* 114 (2003): 660–664; A. K. Jha, M. G. Shlipak, W. Hosmer, C. D. Frances, and W. S. Browner, "Racial Differences in Mortality among Men Hospitalized in the Veterans Affairs Health System," *Journal of the American Medicine Association* 285 (2001): 297–303.

38. W. W. Stead, J. W. Senner, W. T. Reddick, and J. P. Lofgren, "Racial Differences in Susceptibility to Infection by Mycobacterium Tuberculosis," *New England Journal of Medicine* 322 (1990): 422–427.

39. Neil W. Schluger reviews this research in "The Pathogenesis of Tuberculosis: The First One Hundred (and Twenty-Three) Years," *American Journal of Respiratory Cell and Molecular Biology* 32 (2005): 251–256.

40. Michael F. Cantwell, Matthew T. McKenna, Eugene McCray, and Ida M. Onorato, "Tuberculosis and Race/Ethnicity in the United States," *American Journal of Respiratory and Critical Care Medicine* 157 (1998): 1016–1020. See also Marion Torchia, "Tuberculosis among American Negroes: Medical Research on a Racial Disease, 1830–1950," *Journal of the History of Medicine and Allied Sciences* 32 (1977): 252–279.

41. See chapter 5.

42. J. D. C. Bennett and D. G. Stock, "The Longstanding Problem of Flat Feet," *J. R. Army Corps* 135 (1989): 144–146.

43. J. H. Baxter, comp., *Statistics, Medical and Anthropological of the Provost-Marshal-General's Bureau, Derived from Records of the Examination for Military Service during the Late War of the Rebellion of over a Million Recruits, Drafted Men, Substitutes, and Enrolled Men* (Washington: Government Printing Office, 1875), quotations from pp. 161, 237, 176.

44. Lynn T. Staheli, Deanna E. Chew, and Marilyn Corbett, "The Longitudinal Arch: A Survey of Eight Hundred and Eighty-Two Feet in Normal Children and Adults," *Journal of Bone and Joint Surgery* 69 (1987): 426–428, quote on 426.

45. S. F. Stewart, "Human Gait and the Human Foot: An Ethnological Study of Flatfoot, [parts] I [and] II," *Clinical Orthopedics and Related Research* 70 (1970): 111–132.

46. "Hygienic and Physiological Observations," a printed circular that accompanies a letter from Benjamin Woodward to Dear Doctor, 20 August 1863, frames 111–114, reel 1, USSC Papers. On the nineteenth-century notion that therapy had to be adjusted to the patient's race, gender, age, and region, see John Harley

Warner, *The Therapeutic Perspective: Medical Practice, Knowledge, and Identity in America, 1820–1885* (Cambridge, Mass.: Harvard University Press, 1986).

47. Reels 1–12, USSC Papers, contain many references that support this paragraph.

48. B. Woodward, "Report on the Diseases of Colored Troops," frame 898, reel 2, USSC Papers.

49. Richard G. Shaw to J. Cary Whiting, 3 October 1864, reproduced in Berlin, *Freedom*, 641.

Four • Medical Care

1. Charles Edward Briggs to Henry, 8 November 1863, in *Civil War Surgeon in a Colored Regiment*, ed. Walter De Blois Briggs (Berkeley: privately published, 1960), 117.

2. Maj. Gen. N. P. Banks to Maj. C. T. Christenden, 18 July 1864, frame 2694, reel 3, NARA M858.

3. Paul E. Steiner, *Medical History of a Civil War Regiment: Disease in the Sixty-Fifth United States Colored Infantry* (Clayton, Mo.: Institute of Civil War Studies, 1977), 30–33.

4. USSC Report # 1415, 1st Regt USCT near Portsmouth, Va., James Hall Jr. is inspector, 24 September 1863, reel 26, USSC Papers.

5. C. W. Foster to J. G. Foster, 20 December 1864, quoted in Joseph T. Glatthaar, *Forged in Battle: The Civil War Alliance of Black Soldiers and White Officers* (New York: Free Press, 1990), 188.

6. Banks to Christenden, 18 July 1864.

7. Unsigned letter to Sir, 20 August 1864, Brazos Santiago, Texas, Record Group 94, NARA, reprinted in Berlin, *Freedom*, 640.

8. Michael Flannery, *Civil War Pharmacy: A History of Drugs, Drug Supply and Provisions, and Therapeutics for the Union and the Confederacy* (New York: Pharmaceutical Products Press, 2004), 80–90; John Herbert Roper, ed., *Repairing the "March of Mars": The Civil War Diaries of John Samuel Apperson, Hospital Steward in the Stonewall Brigade, 1861–1865* (Macon, Ga.: Macon University Press, 2001).

9. Lorenzo Thomas to E. D. Townsend, 8 December 1863, Letters Sent by Adjutant General Lorenzo Thomas, 30 November 1863–July 7, 1864, E-9, Records of the Adjutant General's Office, 1780s–1917, Record Group 94, NARA.

10. Lorenzo Thomas to Col. G. U. Zeigler, 4 February 1864, in ibid. See also Thomas to Brig. Gen. William Pile, 22 February 1864; Thomas to Brig. Gen. A. S. Chetlain, 3 June 1864; Thomas to Colonel A. B. Morrison, 25 June 1864, all in ibid.

11. Lorenzo Thomas to E. D. Townsend, 8 December 1863, in ibid.

12. Fort Green to the editor, 21 August 1864, published, 24 September 1864, *Christian Recorder*.

13. Robert G. Slawson, "African American Physicians in the Union Army dur-

ing the Civil War," *Journal of Civil War Medicine* 7 (2003): 47–52, identifies these ten. There may be others, especially in the position of acting assistant surgeon, or contract surgeon, that have been lost in the administrative record. See also Robert G. Slawson, *Prologue to Change: African Americans in Medicine in the Civil War Era* (Frederick, MD: NMCWM Press, 2006).

14. A. T. Augusta to President Abraham Lincoln, 7 January 1863, in Berlin, *Freedom*, 354; Order appointing Augusta as surgeon, 4 April 1863, in frame 1171, reel 2, NARA M858.

15. "Important Letter from Dr. Augusta," *Anglo-African*, 13 April 1863.

16. Reprint of a letter from Augusta to the editor of the *Republican*, 15 May 1863, published 30 May 1863, *Christian Recorder*.

17. *Congressional Globe*, 9 February 1864, 38th Cong., 1st Sess., pp. 553–555.

18. "Negro Equality in the North," *Richmond Examiner*, 6 April 1864.

19. J. B. McPherson et al. to President Abraham Lincoln, February 1864, in Berlin, *Freedom*, 356–357.

20. Joel Morse to John Sherman, May 1864, in ibid., 357.

21. Slawson, "African American Physicians"; Douglas M. Haynes, "Policing the Social Boundaries of the American Medical Association (AMA), 1847–1870," *Journal of the History of Medicine and Allied Sciences* 60 (2005): 170–195.

22. John V. Degrasse Papers, Personal Papers of Medical Officers and Physicians, Records of the Adjutant General's Office, 1780s–1917, Record Group 94, NARA; Glatthaar, *Forged in Battle*, 189.

23. Record of contract surgeons, E-575, Records of the Adjutant General's Office, 1780s–1917, Record Group 94, NARA.

24. J. Dennis Harris, *A Summer on the Borders of the Caribbean Sea* (1860); reprinted in full in *Black Separatism and the Caribbean 1860*, ed. Howard H. Bell (Ann Arbor: University of Michigan Press, 1970).

25. J. D. Harris, Autobiography, Personal Papers of Medical Officers and Physicians, Records of the Adjutant General's Office, 1780s–1917, Record Group 94, NARA. Harris describes the Iowa school as the Medical Department of Iowa University, in Keokuk. There are no surviving alumni records (e-mail message from Susan Lawrence, 1 December 2004, who has studied the history of this school). Jill Tatem, acting director of the University Archives at Case Western Reserve University reported that Joseph D. Harris attended the Medical Department of Western Reserve College in 1863–1864 but did not graduate (e-mail message to Robert Slawson, 19 October 2004).

26. LOOK IN to the editor, 2 July 1864, published 9 July 1864, *Christian Recorder*.

27. Lucy Chase to Dear Friends, 1 July 1864, in *Dear Ones at Home: Letters from Contraband Camps*, ed. Henry L. Swint (Nashville: Vanderbilt University Press, 1966), 122–123.

28. Harris Papers, NARA.

29. Ira Russell, "Report on Hospitals in Richmond, Norfolk, etc.," frame 155, reel 3, USSC Papers; W. D. S. to the editor, 7 November 1865, published 23 December 1865, *Christian Recorder.*

30. *Richmond Enquirer and Examiner*, 11 March 1869; *Richmond Whig and Advertiser*, 12 March 1869, A. A. Taylor, "Reconstruction through Compromise," *Journal of Negro History* 11 (1926): 494–512; Richard Low, *Republicans and Reconstruction in Virginia, 1856–70* (Charlottesville: University Press of Virginia, 1991).

31. Slawson, "African American Physicians."

32. J. W. Compton, report, 20 May 1865, in *Statistics, Medical and Anthropological of the Provost-Marshal-General's Bureau, Derived from Records of the Examination for Military Service during the Late War of the Rebellion of over a Million Recruits, Drafted Men, Substitutes, and Enrolled Men*, comp. J. H. Baxter (Washington, D. C.: Government Printing Office, 1875), 368–369.

33. W. F. Johnson to editor, Camp Bailey, Dutch Island, R.I., 5 March 1864, published 12 March 1864, *Anglo-African.*

34. Benjamin Woodward, "Notes on a Trip through Arkansas," frame 558, reel 1, USSC Papers.

35. Benjamin Woodward to Elisha Harris, 19 September 1865, frame 632, reel 2, USSC Papers.

36. Physicians quoted in Glatthaar, *Forged in Battle*, 190.

37. Keith P. Wilson, *Campfires of Freedom: The Camp Life of Black Soldiers during the Civil War* (Kent, Ohio: Kent State University Press, 2002), 109–146.

38. Margaret Humphreys, *Malaria: Poverty, Race, and Public Health in the United States* (Baltimore: Johns Hopkins University Press, 2001), 113–139.

39. Charles E. Rosenberg, "The Therapeutic Revolution: Medicine, Meaning, and Social Change in Nineteenth-Century America," *Perspectives in Biology and Medicine* 20 (1977): 485–506.

40. Benjamin Quarles, *The Negro in the Civil War* (1953; rpt., New York: Da Capo, 1989), 225–226; Susie King Taylor, *A Black Woman's Civil War Memoirs: Reminiscences of My Life in Camp with the 33rd U.S. Colored Troops, Late 1st South Carolina Volunteers*, ed. Patricia W. Romero, introduction by Willie Lee Rose (orig. ed. 1902; rpt., New York: Markus Wiener, 1988).

41. George Andrew to Elisha Harris, 17 July 1865, frame 418, reel 1, USSC Papers.

42. B. Woodward, "Report on the Diseases of Colored Troops," frame 892, reel 2, USSC Papers; Benjamin Woodward to Elisha Harris, 19 September 1865, frame 711, reel 2, USSC Papers. Sharla Fett discusses the magical world of slave health and healing in *Working Cures: Healing, Health, and Power on Southern Slave Plantations.* (Chapel Hill: University of North Carolina Press, 2002).

43. Charles Edward Briggs to Dear Emma, 18 October 1864, in Briggs, *Civil War Surgeon*, 138.

44. Ira Russell, "Report on Hospital L'Ouverture," Alexandria, Va., frames 282–283, reel 3, USSC Papers.

45. Glatthaar, *Forged in Battle*, 191. The surgeons subsequently lost their positions because of these actions.

46. Versalle F. Washington, *Eagles on Their Buttons: A Black Infantry Regiment in the Civil War* (Columbia: University of Missouri Press, 1999), 22. See also Thomas P. Lowry and Jack Welsh, eds., *Tarnished Scalpels: The Court-Martials of Fifty Union Surgeons* (Mechanicsburg, Pa.: Stackpole, 2000), 41.

47. Unsigned letter to Sir, 20 August 1864, Brazos Santiago, Texas, Record Group 94, NARA, in Berlin, *Freedom*, 640. "Buck and gag" referred to a punishment in which the soldier sat on the ground with his hands and feet bound. His knees were drawn up between his arms, and a stick passed under the knees and over the arms to keep him in the folded up position. A stick was placed across the mouth and tied behind the head to form a gag; some particularly cruel officers used a bayonet or knife for this purpose.

48. J. B. McPherson et al. [including E. M. Pease] to President Abraham Lincoln, February 1864, in Berlin, *Freedom*, 356–357; Order creating medical examining board for 25th Army Corps, 4 May 1865, in J. D. Harris Papers, Box 249, Personal Papers of Medical Officers and Physicians, Records of the Adjutant General's Office, 1780s–1917, Record Group 94, NARA. Pease's service record is reviewed at www.itd.nps.gov/cwss (accessed 2 February 2005).

49. B. W. to editor, *Christian Recorder*, 30 July 1864.

50. Benjamin Woodward, "Report on the Diseases of Colored Troops," frame 893, reel 2, USSC Papers.

51. Rufus S. Jones to the editor, 20 March 1864, published 16 April 1864, *Christian Recorder;* Jones to the editor, 13 April 1863, published 7 May 1864, *Christian Recorder.*

52. Ira Russell, "Report on Hospital L'Ouverture," Alexandria, Va, frames 288–289, reel 3, USSC Papers.

53. Quoted in Glatthaar, *Forged in Battle*, 194.

54. Lorenzo Thomas to G. H. Thomas, 27 June 1864, Letters Sent by Adjutant General Lorenzo Thomas, Nov. 30, 1863–July 7, 1864, Records of the Adjutant General's Office, 1780s–1917, Record Group 94, NARA.

55. Joseph K. Barnes to Ira Russell, 17 December 1864; Russell to Barnes, 31 January 1865; J. H. Brinton to Barnes, 12 March 1865, Ira Russell Papers, Personal Papers of Medical Officers and Physicians, Records of the Adjutant General's Office, 1780s–1917, Record Group 94, NARA.

56. Lorenzo Thomas to R. C. Wood, 16 January 1865, in Berlin, *Freedom*, 645.

57. Ira Russell confronted army inertia in St. Louis during the severe disease outbreaks of 1863–1864; see chapter 5. The Nashville story is described above. When scurvy erupted among the black troops in Texas, the small hospital there was inadequate to care for them, and the hundreds shipped to New Orleans rapidly overwhelmed the available hospital space. See chapter 7.

58. Glatthaar, *Forged in Battle*, 193.

59. Jane Schultz, *Women at the Front: Hospital Workers in Civil War America* (Chapel Hill: University of North Carolina Press, 2004), 6.

60. Abby Hopper Gibbons, quoted in ibid., 136–137. The Helena, Arkansas, case is also described in ibid., 137.

61. Ibid., 22 and 33; Jane Schultz, "Seldom Thanked, Never Praised, and Scarcely Recognized: Gender and Racism in Civil War Hospitals," *Civil War History* 48 (2002): 220–236.

62. Taylor, *A Black Woman's Civil War.*

63. Quarles, *The Negro in the Civil War,* 226, 228.

64. Schultz, *Women at the Front,* 98–104.

Five · Region, Disease, and the Vulnerable Recruit

1. Thomas Wentworth Higginson, *Army Life in a Black Regiment* (1870), with introduction by Howard Mumford Jones (East Lansing: Michigan State University Press, 1960), 27.

2. "Endurance of the Black Soldiers," *Weekly Anglo-African,* 5 September 1863.

3. James C. Beecher to Edward A. Wild, 13 September 1863, in Berlin, *Freedom,* 493.

4. R. T. Auchmuty to E. D. Townsend, 20 December 1863, frames 1854–1855, reel 3, NARA M858.

5. Concepts of disease causation in the 1860s were in a state of flux. Increasingly, physicians believed that specific agents—poisons, ferments, or perhaps fungi—spread infectious diseases. As examples of such thinking, see Elisha Harris, "Control and Prevention of Infectious diseases," in *Military Medical and Surgical Essays Prepared for the United States Sanitary Commission,* ed. William A. Hammond (Philadelphia: J. B. Lippincott and Co., 1864), 45–90, and Joseph Janvier Woodward, *Outlines of the Chief Camp Diseases of the United States Armies as Observed during the Present War: A Practical Contribution to Military Medicine* (1863; rpt., New York: Hafner, 1964). Woodward doubted the influence of such microscopic substances but in the process of discussing them reviews many of the contending ideas of his day. The best secondary source on analyzing this confusing period is Michael Worboys, *Spreading Germs: Disease Theories and Medical Practice in Britain, 1865–1900* (Cambridge: Cambridge University Press, 2000).

6. Calculated from information in Frederick H. Dyer, *A Compendium of the War of the Rebellion* (New York: Thomas Youseloff, 1959), 3:1720–1740. These numbers are necessarily loose, as some regiments were mixed. The regiments raised in Boston, New York, and Philadelphia, as well as Connecticut and Michigan, were composed of free men, although some may have been born in slavery and escaped. Eight of the seventeen regiments that saw duty in South Carolina come from one of these locations. Two were from Maryland and probably included slaves liberated from farms specifically for service in the army, although others

may have been free. A similar ambiguity surrounds regiments raised in Kentucky. For regiments active in Virginia, fourteen out of thirty-two came from nonslave states. Another twelve came from Kentucky and Maryland and were likely mixed in terms of prior legal status. In North Carolina, three of nine regiments came from nonslave states, one from the District of Columbia, and two from border states.

7. Robert Gould Shaw to father, 21 March and 2 April 1863, in *Blue-Eyed Child of Fortune: The Civil War Letters of Colonel Robert Gould Shaw*, ed. Russell Duncan (Athens: University of Georgia Press, 1992), 311 and 319.

8. J. H. Baxter, comp. *Statistics, Medical and Anthropological of the Provost-Marshal-General's Bureau, Derived from Records of the Examination for Military Service during the Late War of the Rebellion of over a Million Recruits, Drafted Men, Substitutes, and Enrolled Men* (Washington, D.C.: Government Printing Office, 1875), 1:1–360.

9. C. W. Foster to Col. S. Bowman, 17 June 1864, Colored Troops Division, 1863–1889, Letters Sent, Dec 1863–March 1888, vol. 2a, E. 352, Records of the Adjutant General's Office, 1780s–1917, Record Group 94, NARA.

10. Louis S. Gerteis, *Civil War St. Louis* (Lawrence: University Press of Kansas, 2001); William C. Winter, *The Civil War in St. Louis* (St. Louis: Civil War Round Table and Missouri Historical Society Press, 1994).

11. Dena Lange, *A History of St. Louis: The City Surrounded by the United States* (St. Louis: Public School Messenger, 1931), 2:243. These were American Sisters of Charity derived from the order begun by Elizabeth Ann Seton in Emmitsburg, Maryland, and they arrived in St. Louis in 1828.

12. E. J. Goodwin, *A History of Medicine in Missouri* (St. Louis: W. L. Smith, 1905).

13. Gerteis, *Civil War St. Louis*, 202–219; Robert Patrick Bender, "Old Boss Devil: Sectionalism, Charity, and the Rivalry between the Western Sanitary Commission and the United States Sanitary Commission during the Civil War," Ph.D. diss., University of Arkansas, 2001.

14. Galusha Anderson, *The Story of a Border City during the Civil War* (Boston: Little, Brown, and Co., 1908), 251.

15. Ibid., 256, 261.

16. Charles Van Ravenswaay, *Saint Louis: An Informal History of the City and Its People, 1764–1865* (St. Louis: Missouri Historical Society, 1991), 498–99. A useful guide to the history of Benton Barracks, with photographs and maps, is *http://missouricivilwarmuseum.org/benton.htm* (accessed 10 March 2005).

17. F. F. Kiner, *One Year's Soldiering* (1863; rpt., Prior Lake, Minn.: Morgan Ave. Press, 2000), 6.

18. D. L. McGugin, "Sanitary Condition of Benton Barracks, St. Louis, Missouri, 1861–62," *Medical and Surgical Reporter* 12 (1865): 379–380, 473–475, quotations on 379 and 474.

19. Ibid., 474.

20. Kiner, *One Year's Soldiering*, 6.

21. McGugin, "Sanitary Condition," 474–475.

22. *The Western Sanitary Commission; A Sketch of Its Origin, History, Labors for the Sick and Wounded of the Western Armies, and Aid Given to Freedmen and Union Refugees, with Incidents of Hospital Life* (St. Louis: R. P. Studley and Co., 1864), 73.

23. See multiple documents in Ira Russell Papers, Personal Papers of Medical Officers and Physicians, Records of the Adjutant General's Office, 1780s–1917, Record Group 94, NARA. Other details are found in "Ira Russell," in *Physicians and Surgeons of the United States,* ed. William Atkinson (Philadelphia: C. Robson, 1878), 310–311.

24. Ira Russell to Dear Wife, 20 February 1863, in Ira Russell Papers, Collection 4440, Southern Historical Collection, Wilson Library, University of North Carolina at Chapel Hill.

25. E. L. Phillips to Ira Russell, 1 March 1863, in Russell Papers, UNC Chapel Hill.

26. Emily Parsons, quoted in Theophilus Parsons, ed., *Memoir of Emily Elizabeth Parsons* (Boston: Little, Brown, and Co., 1880), 72, 90, 95. This volume mainly consists of Parsons's letters, with an introduction by Theophilus, her father.

27. Letters from August and September 1863 in Russell Papers, NARA.

28. Parsons, *Memoir,* 133.

29. Ira Russell, "The Sanitary Report of Benton Barracks near St Louis, Missouri to the United States Sanitary Commission," frames 251–253, reel 2, USSC Papers.

30. Ira Russell to Henry Wilson, 20 December 1863, frames 778–780, reel 10, USSC Papers.

31. General Order 37, 26 December 1862, *OR,* ser. 1, vol. 22, pt. 1, p. 876, includes Mills's appointment as Medical Director of the Department of the Missouri.

32. William F. Fox, *Regimental Losses in the American Civil War, 1861–1865,* 4th ed. (Albany, N.Y.: Joseph McDonough, 1898), 524.

33. Russell to Wilson, 20 December 1863.

34. Russell, "The Sanitary Report of Benton Barracks near St Louis, Missouri to the United States Sanitary Commission," frame 253, reel 2, USSC Papers.

35. [Ira Russell to Sir], 31 January 1864, folder 2, Russell Papers, UNC Chapel Hill.

36. Madison Mills to Joseph Barnes, 18 March 1864, in Russell Papers, NARA. Russell's service record is also contained in this collection.

37. Dyer, *Compendium,* 3:1733–34; Paul E. Steiner, *Medical History of a Civil War Regiment: Disease in the Sixty-Fifth United States Colored Infantry* (Clayton, Mo.: Institute of Civil War Studies, 1977). 6. The *Sultana* is best remembered for exploding while transporting liberated prisoners from Andersonville in 1865; see

Lonnie R. Speer, *Portals to Hell: Military Prisons of the Civil War* (Mechanicsburg: Stackpole Books, 1997), 289.

38. Margaret Humphreys, "A Stranger in Our Camps: Typhus in American History," *Bulletin of the History of Medicine* 80 (2006): 269–290.

39. Steiner, *Medical History*, 84; Philip A. Brunell, "Measles (*Morbilli, Rubeola*)," in *Cecil Textbook on Medicine*, 19th ed., ed. James B. Wyngaarden, Lloyd H. Smith, and J. Claude Bennett (Philadelphia: W. B. Saunders, 1992), 1825–1827.

40. Ira Russell, "Pneumonia as It Appeared among the Colored Troops at Benton Barracks, Mo., during the Winter of 1864," in *Contributions Relating to the Causation and Prevention of Disease and to Camp Diseases; Together with a Report of the Diseases, Etc., among the Prisoners at Andersonville, Ga.*, ed. Austin Flint (New York: Hurd and Houghton, 1867), 319–334, quote on p. 332.

41. Ira Russell, "Report on Spurious or Impure Vaccination," frame 345, reel 4, USSC Papers.

42. Margaret Schibuk, "The Search for Vaccinia," Ph.D. diss., Harvard University, 1986.

43. Ira Russell, quoted in S. B. Hunt, "Spurious Vaccination," frames 278–280, reel 5, USSC Papers.

44. [A.?] Hammer, quoted in James W. Clemens, "Proceedings of the St. Louis Medical Society: A Discussion of Vaccination," *St. Louis Medical and Surgical Journal* 2 (1865): 320–332, quote on p. 328; *MSHW*, 6:637.

45. *MSHW*, 6:635–636.

46. Joseph R. Smith, Medical Director, "Sanitary Report of the Department of Arkansas for the Year 1864," MS C 126, Historical Collections, National Library of Medicine, Bethesda, Md.

47. *MSHW*, 6:625, 628–629.

48. Steiner, *Medical History*, xvi and 99.

49. *MSHW*, 1:658–659; 6:627

50. Ira Russell, "Cerebro Spinal Meningitis, as It Appeared among the Colored Soldiers at Benton Barracks, Mo., during the Winter of 1863–4," *St. Louis Medical and Surgical Journal* new ser. 1 (1864): 121–128. Nine meningitis cases and eight deaths occurred among the 65th USCI (Steiner, *Medical History*, xvi, 57).

51. Ira Russell, "Cerebro-spinal Meningitis as It Appeared among the Troops Stationed at Benton Barracks, Mo.," *Boston Medical and Surgical Journal* 70 (1864): 309–313.

52. *MSHW*, 1:454–455 and 666–667.

53. *MSHW*, 1:552–590; R. E. Houghton, "Epidemic Cerebro-spinal Meningitis: Cerebro-spinal Typhus, Spotted Fever," *Transactions of the Indiana Medical Society* 15 (1865): 47–56; J. Adams Allen, "Some Account of the Prevailing Epidemic in the North-West, Variously Designated, but Usually Popularly Denominated 'Spotted Fever,'" *Chicago Medical Journal* 21(1864): 241–258; Report of a Committee of the Massachusetts Medical Society on Spotted Fever or Cerebro-spinal

Meningitis in the State of Massachusetts, *Publications of the Massachusetts Medical Society* 2 (1868): 1–150; J. J. Levick, "Report of the Committee on 'Spotted Fever, So-called,'" *Transactions of the American Medical Association* 17 (1866): 311–364.

54. Sanford B. Hunt, "On Cerebro-Spinal Meningitis," in Flint, *Contributions Relating to the Causation and Prevention of Disease*, 383–411.

55. See *MSHW*, 1:587–588, on Giesburg, Maryland; Mobile, Alabama; and Grenada, Mississippi. Also on the Alabama and Mississippi outbreaks, see G. A. Moses, "Epidemic Cerebro-Spinal Meningitis," *Confederate States Medical and Surgical Journal* 1 (1864): 113–115. On the condition of slave workers on the fortifications at Richmond, see William Carrington to Surg. George Vest, Engineering Hospital, 21 January 1865, chapter 6, vol. 416, Letters Sent, Medical Director's Office, Richmond, Va., Record Group 109, War Department Collection of Confederate Records, NARA; and R. P. Vest, "Epidemic Cerebro-spinal Meningitis," *Richmond Medical Journal* 1 (1866): 314–318.

56. Michael Sappol, *A Traffic of Dead Bodies: Anatomy and Embodied Social Identity in Nineteenth-Century America* (Princeton: Princeton University Press, 2002); Ira Russell to Medical Committee, 15 July 1865, frame 969, reel 1, USSC Papers.

57. Harriet Martineau, *Retrospect of Western Travel* (London: Saunders and Otley, 1838), 140.

58. David C. Humphrey, "Dissection and Discrimination: The Social Origins of Cadavers in America, 1760–1915," *Bulletin of the New York Academy of Medicine* 49 (1973): 819–827.

59. Sappol, *Traffic*, 123–124; "Ira Russell," in Atkinson, *Physicians and Surgeons of the United States*, 310–311.

60. These records are in frames 200–269, reel 5, USSC Papers. The last page with text is frame 252; 253–269 are blank pages, with the ruled grid prepared, with space for another 208 entries.

Six · Louisiana

1. James M. McPherson, *Battle Cry of Freedom: The Civil War Era* (New York: Ballantine Books, 1988), 637.

2. Abraham Lincoln quoted in ibid., 638.

3. On the reputation for disease of the lower Mississippi Valley, see Margaret H[umphreys] Warner, "Public Health in the Old South," in *Science and Medicine in the Old South*, ed. Ronald L. Numbers and Todd L. Savitt (Baton Rouge: Louisiana State University Press, 1989), 226–255. On yellow fever and malaria, see Margaret Humphreys, *Malaria: Poverty, Race and Public Health in the United States* (Baltimore: Johns Hopkins University Press, 2001), and *Yellow Fever and the South* (New Brunswick, N.J.: Rutgers University Press, 1992).

4. S. B. Holabird to Maj. Gen. Banks, 22 Dec 1862, Field Records of the Banks Expedition, Department of the Gulf, RG 393, NARA, in Berlin, *Freedom*, 638.

5. H. W. Halleck to U. S. Grant, 31 March 1863, H. W. Halleck Letters Sent, RG 94, NARA, reprinted in Berlin, *Freedom*, 143.

6. Data compiled from regimental entries in Frederick H. Dyer, *A Compendium of the War of the Rebellion* (New York: Thomas Youseloff, 1959), 3:1720–1740.

7. *MSHW*, 1:246–251, 396–401, and 548–553.

8. *Richmond Examiner*, 25 February 1864, 1.

9. James E. Yeatman, *A Report on the Condition of the Freedmen of the Mississippi, Presented to the Western Sanitary Commission, December 17th, 1863* (Saint Louis: Western Sanitary Commission, 1864), 12.

10. Gaines M. Foster, "The Limitations of Federal Health Care for Freedmen, 1862–1868," *Journal of Southern History* 48 (1982): 349–372.

11. William F. Messner, "The Federal Army and Blacks in the Gulf Department, 1862–1865" (Ph.D. diss., University of Wisconsin, 1972), 329.

12. Frederick Phisterer, *Statistical Record of the Armies of the United States: Campaigns of the Civil War, Supplementary Volume 13* (New York: Charles Scribner's Sons, 1901), 215; Lawrence Lee Hewitt, "An Ironic Road to Glory: Louisiana's Native Guards at Port Hudson," in *Black Soldiers in Blue: African American Troops in the Civil War Era*, ed. John David Smith (Chapel Hill: University of North Carolina Press, 2002), 78–106.

13. Henry Wilson, 6 June 1864, *Congressional Globe*, 38th Congress, First Session, p. 2766.

14. Report of G. W. Avery about 81st Colored Infantry, frame 700, reel 1, USSC papers.

15. George G. Edgerly, inspection reports to the U.S. Sanitary Commission, October 23–November 7, 1863, frames 1421–1447, reel 26, USSC Papers.

16. J[ames] B. N[ickerson] to Dear Quincy, Port Hudson, La, 24 April 1864, Civil War Papers, Dept. of Special Collections, Stanford University Libraries, Stanford, California. On the Fort Pillow massacre, see Andrew Ward, *River Run Red: The Fort Pillow Massacre in the American Civil War* (New York: Viking, 2005).

17. Richard S. Offenberg and Robert Rue Parsonage, eds., *The War Letters of Duren F. Kelley, 1862–1865* (New York: Pageant, 1967).

18. Messner, "The Federal Army and Blacks in the Gulf," 350.

19. Offenberg and Parsonage, *War Letters*.

20. Charles Dickens, *Martin Chuzzlewit*, edited with an introduction by Margaret Cardwell (Oxford: Clarendon, 1982), 380.

21. Duren Kelley to Emma Kelley, 14 March 1865, in Offenberg and Parsonage, *War Letters*, 150.

22. Duren Kelley to Emma Kelley, 27 March 1864, in ibid., 96.

23. Duren Kelley to Emma Kelley, 5 May 1864, in ibid., 97.

24. Duren Kelley to Emma Kelley, 28 May 1864, in ibid., 101.

25. Duren Kelley to Emma Kelley, 8 June 1864, in ibid., 103.

26. Paul E. Steiner, *Medical History of a Civil War Regiment: Disease in the Sixty-Fifth United States Colored Infantry* (Clayton, Mo.: Institute of Civil War Studies, 1977), 9. This place is also called Morganza Bend, or Morganzia. Morganza is the modern spelling.

27. Duren Kelley to Emma Kelley, 4 July 1864, in Offenberg and Parsonage, *War Letters*, 105.

28. Duren Kelley to Emma Kelley, 7 July 1864, in ibid., 109.

29. Duren Kelley to Emma Kelley, 24 January 1865, in ibid., 136.

30. Duren Kelley to Emma Kelley, 7 July 1864, in ibid., 109.

31. Jno. L. Rice to L. Thomas, 10 Sept 1864, RG 94, NARA, in Berlin, *Freedom*, 509.

32. Duren Kelley to Emma Kelley, 24 July 1864, in Offenberg and Parsonage, *War Letters*, 111.

33. Nimrod Rowley to Abraham Lincoln, [Aug.] 1864, RG 94, NARA, in Berlin, *Freedom*, 502.

34. Daniel Ullmann to C. T. Christensen, 29 Oct 1864, RG 393, NARA, in ibid., 513.

35. Keith P. Wilson, *Campfires of Freedom: The Camp Life of Black Soldiers during the Civil War* (Kent, Ohio: Kent State University Press, 2002), 40.

36. John Cajay to the editor, 7 September 1864, published 15 October 1864, *Weekly Anglo African*.

37. Ibid.

38. H. N. Frisbie to O. A. Rice, 24 Sept. 1864, R.G. 94, NARA, in Berlin, *Freedom*, 510.

39. Duren Kelley to Emma Kelley, September 1864, in Parsonage and Offenberger, *War Letters*, 117.

40. Army inspector quoted in Joseph Glatthaar, *Forged in Battle: The Civil War Alliance of Black Soldiers and White Officers* (Baton Rouge: Louisiana State University Press, 1990), 193.

41. Table accompanying letter from Daniel Ullmann to C. T. Christenson, 29 Oct 1864, RG 393, NARA, in Berlin, *Freedom*, 514.

42. Vincent Cirillo, *Bullets and Bacilli: The Spanish-American War and Military Medicine* (New Brunswick, N.J.: Rutgers University Press, 2004).

Seven • Death on the Rio Grande

1. Ira Perry, "Case 52—Private Noah Davis," in *MSHW*, 6:701. Noah Davis's admission and death are recorded in the Register of Patients at Brownsville Post Hospital, Arranged Alphabetically, Sept. 1865 to April 1866, vol. 9, entry 544, Field Records of Hospitals, Records of the Adjutant General's Office, 1780s–1917,

Record Group 94, NARA. I have taken some liberties here in recording what Davis thought, giving him reactions and opinions held by fellow soldiers who lived to record their thoughts and are cited later in this chapter.

2. Jeffrey Hunt, *The Last Battle of the Civil War: Palmetto Ranch* (Austin: University of Texas Press, 2002).

3. See the chapters on the 1850s and 1860s in Michael C. Meyer, William L. Sherman, and Susan M. Deeds, *The Course of Mexican History* (New York: Oxford University Press, 1999), and Michael C. Meyer and William H. Beezley, *The Oxford History of Mexico* (New York: Oxford University Press, 2000).

4. William Robert Fortschen, "The Twenty-Eighth United States Colored Troops: Indiana's African Americans Go to War, 1863–1865," Ph.D. dissertation, Purdue University, 1994; Berlin, *Freedom*, 734. The regimental count is compiled from Frederick H. Dyer, *A Compendium of the War of the Rebellion* (New York: Thomas Youseloff, 1959), 3:1720–1740.

5. Alexander Herritage [*sic*] Newton, *Out of the Briars: An Autobiography and Sketch of the Twenty-Ninth Regiment Connecticut Volunteers* (Philadelphia: AME Book Concern, 1910), 69–70.

6. Alexander McDonald to Elisha Harris, 2 July 1865, frame 470, reel 2, USSC Papers.

For a modern description, see *www.tsha.utexas.edu/handbook/online/articles/BB/rrb10.html* (accessed 25 August 2005). The fort buildings were destroyed by a hurricane in 1867.

7. M. R. Williams to the editor, Brazos Santiago, Texas, 30 June 1865, published 29 July 1865, *Christian Recorder.*

8. S[amuel] H. Smothers to the editor, 24 June 1865, published 15 July 1865, *Christian Recorder.*

9. James D. Birkett, "The 1861 de Normandy Desalting Unit at Key West," *Desalination and Water Reuse Quarterly* 7(1997): 53–57.

10. Charles W. Cole to editor, Brownsville, Texas, 20 August 1865, published 9 September 1865, *Christian Recorder.*

11. Newton, *Out of the Briars,* 69–70.

12. "Rufus" to editor, Indianola, Texas, 15 July 1865, published in the *Weekly Afro-American,* 12 August 1865, reprinted in Edwin S. Redkey, ed., *A Grand Army of Black Men: Letters from African-American Soldiers in the Union Army, 1861–1865* (Cambridge: Cambridge University Press, 1992), 197–198, quote on p. 198. The Croton aqueduct brought water into New York City.

13. Williams to the editor, *Christian Recorder,* 30 June 1865.

14. Dr. McDonald's "Report on the Texas Expedition and 25th Army Corps," 30 October 1865, frame 697, reel 6, USSC Papers.

15. Garland H. White, Chaplain, 28th USCI, Corpus Christi, Texas, to editor, 19 September 1865, published 21 October 1865, *Christian Recorder.*

16. Newton, *Out of the Briars,* 80.

17. Williams to the editor, *Christian Recorder*, 30 June 1865.

18. S. Hemenway, "Observations on Scurvy and Its Causes among the U.S. Colored Troops of the 25th Army Corps, during the Spring and Summer of 1865," *Chicago Medical Examiner* 7 (1866): 582–587.

19. Charles Smart, "Scurvy," *MSHW*, 6:694.

20. Register of Death, Post Hospital, Brownsville, Texas, vol. 8, entry 544; Brownsville Hospital Register, July to November 16, 1865, Texas, vol. 5, Field Records of Hospitals, RG 94, NARA.

21. White to the editor, *Christian Recorder*, 19 September 1865.

22. Kenneth J. Carpenter, *The History of Scurvy and Vitamin C* (Cambridge: Cambridge University Press, 1986); A. J. Bollet, "Scurvy and Chronic Diarrhea in Civil War Troops: Were They Both Nutritional Deficiency Syndromes?" *Journal of the History of Medicine and Allied Sciences* 47 (1992): 49–67.

23. Dr. McDonald's "Report on the Texas Expedition and 25th Army Corps," 30 October 1865, frame 682, reel 6, USSC Papers.

24. See multiple references to scurvy relief in the USSC papers.

25. Lisa M. Brand, "The Wilderness of War: Nature and Strategy in the American Civil War," *Environmental History* 10 (2005): 421–447.

26. See relevant sections (on Vicksburg, Sherman's campaign in Georgia, and the battles in East Tennessee) in James McPherson, *Battle Cry of Freedom: The Civil War Era* (New York: Ballantine Books, 1988). The USSC papers report on these and other small scurvy outbreaks, and the USSC response. See, particularly, E. B. Wolcott's description of his attempts to get into Chattanooga to inspect conditions there. He crossed parts of the Cumberland Plateau on foot, as the army refused him a train pass, and his baggage was stolen by the Rebels (Report of 5 November 1863, frame 892–898, reel 4, USSC Papers).

27. James A. Irby, *Backdoor at Bagdad: The Civil War on the Rio Grande*, Southwestern Studies no. 53 (El Paso: University of Texas at El Paso, Texas Western Press, 1977), 23.

28. Col. Henry M Day, 91st Illinois Infantry, to Maj. George B Drake, Assistant Adjutant General, Dept of the Gulf, 15 August 1864, Brazos Santiago, in *OR*, ser. 1, vol. 41, part 1, p. 212.

29. Unsigned soldier to Sir, 20 August 1864, from Brazos Santiago, Texas, 20 August, 1864, in Berlin, *Freedom*, 640.

30. James Otis Moore to my dear Lizzie (his wife), 16 June 1865, James Otis Moore Papers, 1822–1886, Special Collections Library, Duke University, Durham, North Carolina.

31. Williams to the *Christian Recorder*, 30 June 1865. Soldiers should not have been foraging at all—the country was at peace and unrecompensed raiding of local farmhouses was thievery not to be justified by a wartime situation.

32. Stephen Hathaway to Cousin Gus, 19 July 1865, Brazos Santiago, Stephen F. Hathaway Correspondence, 1863–1866, Special Collections, Duke University.

33. Moore to Lizzie, 30 June 1865, Moore Papers.

34. "Rufus" to the *Weekly Anglo-African*, 15 July 1865.

35. E. M. Pease to E. P. Vollum, Bvt. Lt. Col. and Medical Director, Dept Texas, Brownsville, Texas, 27 April 1866, in Charles Radmore Papers, Personal Papers of Medical Officers and Physicians, Records of the Adjutant General's Office, 1780s–1917, Record Group 94, NARA.

36. Alexander McDonald to Elisha Harris, 21 July 1865, frame 479, reel 1, USSC Papers.

37. Elisha Harris to Standing Commission of the USSC, New Orleans, 16 July 1865, frame 957, reel 1, USSC Papers.

38. Elisha Harris to C. R. Agnew, 15 July 1865, frame 964, reel 1, USSC Papers.

39. Elisha Harris to C. R. Agnew, 22 July 1865, frame 931, reel 1, USSC Papers.

40. William Quentin Maxwell, *Lincoln's Fifth Wheel: The Political History of the U.S. Sanitary Commission* (New York: Longmans, Green and Co., 1956).

41. Elisha Harris to C. R. Agnew, 22 June 1865, frame 74, reel 1, USSC Papers.

42. Ibid.

43. Elisha Harris to C. R. Agnew, 23 June 1865, frame 74, reel 1, USSC Papers.

44. Dr. McDonald's "Report," frame 682–97, reel 1, USSC Papers.

45. Dyer, *Compendium*, 1720–1740.

46. Williams to the editor, *Christian Recorder*, 30 June 1865.

47. William R. Miller, Sgt. Major, 22nd USCT, to the editor, 24 August 1865, published 23 September 1865, *Christian Recorder*.

48. See the Civil War Soldier and Sailor System website, maintained by the National Park Service: *www.itd.nps.gov/cwss/* (accessed 29 August 2005).

49. Llewellyn F. Haskell to Capt. R. C. Shannon, Edinburg, Texas, 18 September 1865, letter no. A321, Letters of the Colored Troops Division, 1865, box 112, Papers of the Adjutant General's Office, RG 94, NARA. Weitzel's comments are written on the back of the letter in the form of an "endorsement." As a letter or report was passed up the hierarchy, each man receiving it would write a comment on the back that acknowledged receipt and at times offered an opinion.

50. See Newton, *Out of the Briars*. Newton was also the son-in-law of the editor of the *African American Weekly*, the other major black periodical of the time.

51. George G. Potts Papers, Personal Papers of Medical Officers and Physicians, Records of the Adjutant General's Office, 1780s–1917, Record Group 94, NARA. On the first court-martial relating to the anatomical dissection gone awry, see Thomas P. Lowry and Jack D. Welsh, *Tarnished Scalpels: The Court-Martials of Fifty Union Surgeons* (Mechanicsburg, Pa.: Stackpole, 2000), 68–75.

52. George G. Potts to J. K. Barnes, 20 July 1865, Potts Papers.

53. Ibid.

54. Potts to Barnes, 2 September 1865, Potts Papers. A printed version of the 18 October 1865 court-martial and its outcome is also included in these papers.

55. Edward P. Vollum to General, 8 January 1866, Radmore Papers.

56. E. M. Pease to Edward P. Vollum, 27 April 1866, Radmore Papers.

57. Ibid.

58. Letters from Charles Radmore reporting that he was on duty as ordered appear in the Radmore Papers until March 1867.

59. Cha[rle]s Greenleaf wrote to the secretary of American board of foreign missions on 16 January 1892: "Rev Edmund Morris Pease is reported to be a missionary in the Marshall Islands recently" (E. M. Pease Papers, Personal Papers of Medical Officers and Physicians, Records of the Adjutant General's Office, 1780s–1917, Record Group 94, NARA).

60. Hemenway, "Observations on Scurvy."

61. Ibid., 583.

62. Ibid., 585.

63. S. B. Hunt, "The Influence of Scorbutic Diathesis on Surgical Accidents," frame 403, reel 10, USSC Papers.

64. Hunt wrote the chapter on scurvy for the USSC's medical history of the war but toned down his criticism of the army in his published account of the Texas disaster. Sanford B. Hunt, "Scurvy in its Medical Aspect," in *Contributions Relating to the Causation and Prevention of Disease and to Camp Diseases; Together with a Report of the Diseases, Etc., among the Prisoners at Andersonville, Ga.*, ed. Austin Flint (New York: Hurd and Houghton, 1867), 276–290.

65. Smart, "Scurvy," 713–714.

66. Harris to the Standing Commission.

Eight · Telling the Story

1. Ira Russell, "Pneumonia as It Appeared among the Colored Troops at Benton Barracks, Mo., during the Winter of 1864," in *Contributions Relating to the Causation and Prevention of Disease and to Camp Diseases; Together with a Report of the Diseases, Etc., among the Prisoners at Andersonville, Ga.*, ed. Austin Flint (New York: Hurd and Houghton, 1867), 319–334, quote p. 319.

2. Ibid., 322.

3. Ibid., 320.

4. Ibid., 322.

5. Ibid., 332.

6. Ibid., 333.

7. Ira Russell, "Cerebro-spinal Meningitis as It Appeared among the Troops Stationed at Benton Barracks, Mo.," *Boston Medical and Surgical Journal* 70 (1864): 309–313, quote on p. 309.

8. "Ira Russell," in *Physicians and Surgeons of the United States*, ed. William Atkinson (Philadelphia: Charles Robson, 1878), 310–311; biography of Russell accompanying website of Ira Russell Letters, Special Collections, University of Arkansas, Fayetteville, *www.uark.edu/libinfo/speccoll/irarussellaid.html* (accessed 22

September 2004). For a listing of Russell's later writings on insanity, see the National Library of Medicine's Indexcat, an online catalogue that includes nineteenth-century medical writings.

9. S. Hemenway, "Observations on Scurvy and Its Causes among the U.S. Colored Troops of the 25th Army Corps, during the Spring and Summer of 1865," *Chicago Medical Examiner* 7 (1866): 585.

10. James B. Fry, "Report of the Provost-Marshal-General's Office," 17 March 1866, in *OR*, ser. 3, vol. 5, 599–932, quote on p. 669.

11. Ibid.

12. Edward S. Dunster, "The Comparative Mortality in Armies from Wounds and Disease," in Flint, *Contributions Relating to the Causation and Prevention of Disease*, 169–192, quote on p. 184.

13. Roberts Bartholow, "The Various Influences Affecting the Physical Endurance, the Power of Resisting Disease, etc., of the Men Composing the Volunteer Armies of the United States," in Flint, *Contributions Relating to the Causation and Prevention of Disease*, 3–41. Bartholow's war service is described in James Holland, "Memoir of Roberts Bartholow, M.D.," *Transactions of the College of Physicians of Philadelphia*, 3d ser., 26 (1904): 43–52.

14. J. H. Baxter, comp. *Statistics, Medical and Anthropological of the Provost-Marshal-General's Bureau, Derived from Records of the Examination for Military Service during the Late War of the Rebellion of over a Million Recruits, Drafted Men, Substitutes, and Enrolled Men* (Washington, D.C.: Government Printing Office, 1875), 161.

15. Roberts Bartholow, *A Manual for Enlisting and Discharging Soldiers, with Special Reference to the Medical Examination of Recruits, and the Detection of Disqualifying and Feigned Diseases* (1863; rpt, San Francisco: Norman Publishing, 1991).

16. Baxter, *Statistics*, 162.

17. Calculations by the author, from Baxter, *Statistics*. All questionnaires were returned from Massachusetts, Connecticut, Rhode Island, New Hampshire, Vermont, and Maine. New York, New Jersey, Pennsylvania, Delaware, and Maryland returned 59%. Ohio, Indiana, Kentucky, Missouri, Illinois, Iowa, Michigan, Wisconsin, and Minnesota returned 51%.

18. Baxter, *Statistics*, 170.

19. Ibid., 215.

20. Ibid., 280.

21. Ibid., 285.

22. Ibid., 461.

23. Ibid., 199.

24. Ibid., 418.

25. Bartholow, *Manual for Enlisting and Discharging Soldiers*, 205.

26. Baxter, *Statistics*, 342.

27. Ibid., 31 and 45. See charts by number; there are no page numbers on chart pages.

28. Ibid.; see charts.

29. Sanford B. Hunt, "The Negro as Soldier," *Anthropological Review* 7 (1869): 40–54, quote on p. 40.

30. Ibid., 43.

31. Ibid.

32. Ibid., 44–48.

33. Ibid., 49

34. Ibid., 51.

35. Ibid.

36. Ibid., 53.

37. Benjamin Apthorp Gould, *Investigations of the Military and Anthropological Statistics of American Soldiers* (Cambridge: Riverside, 1869).

38. John S. Haller Jr., *Outcasts from Evolution: Scientific Attitudes of Racial Inferiority, 1859–1900* (Urbana: University of Illinois Press, 1971); Gould, *Investigations*, 348.

39. Gould, *Investigations*, 319.

40. Cited and discussed in Lundy Braun, "Spirometry, Measurement, and Race in the Nineteenth Century," *Journal of the History of Medicine and Allied Sciences* 60 (2005): 135–169.

41. Ibid.

42. Stephen Jay Gould, *The Mismeasure of Man* (New York: W. W. Norton, 1981).

43. Haller, *Outcasts from Evolution*, 34.

Epilogue

1. David R. Shaffer, *After the Glory: The Struggles of Black Civil War Veterans* (Kent, Ohio: Kent State University Press, 2004).

2. Chulhee Lee, "Wealth Accumulation and the Health of Union Army Veterans, 1860–1870," *Journal of Economic History* 65 (2005): 352–385; quotation on p. 368.

3. Margaret Humphreys, Truls Østbye, Kerry L. Haynie, Idrissa Boly, Philip Costanzo, and Frank Sloan, "Racial Disparities in Diabetes a Century Ago: Evidence from the Pension Files of U.S. Civil War Veterans," *Social Science and Medicine* 64 (2007): 1766–1775.

4. Dora Costa and Matthew E. Kahn, "Forging a New Identity: The Costs and Benefits of Diversity in Civil War Combat Units for Black Slaves and Freemen," *Journal of Economic History* 66 (2006): 936–962.

5. Douglas M. Haynes, "Policing the Social Boundaries of the American Medical Association, 1847–70," *Journal of the History of Medicine and Allied Sciences* 60 (2005): 170–195. There is a much more detailed discussion of the fate of black troops after the war in Joseph Glatthaar, *Forged in Battle: The Civil War Alliance of*

Black Soldiers and White Officers (Baton Rouge: Louisiana State University Press, 1990), 231–264, and in Shaffer, *After the Glory*.

6. *Richmond Enquirer and Examiner*, 11 March 1869; *Richmond Whig and Advertiser*, 12 March 1869; A. A. Taylor, "Reconstruction through Compromise," *Journal of Negro History* 11 (1926): 494–512; Richard Low, *Republicans and Reconstruction in Virginia, 1856–70* (Charlottesville: University Press of Virginia, 1991).

7. Shaffer, *After the Glory*, 48–49.

8. Dora Costa, "Race and Older Age Mortality: Evidence from Union Army Veterans," revised version of NBER Working Paper No. 10902 available at http://mit.edu/costa/www/bwmort5.pdf (accessed 9 August 2006).

9. See, e.g., David H. Donald, ed., *Why the North Won the Civil War* (Baton Rouge: Louisiana State University Press, 1960); Richard E. Beringer, *Why the South Lost the Civil War* (Athens: University of Georgia Press, 1986); and David J. Eicher, *Dixie Betrayed: How the South Really Lost the Civil War* (Boston: Little, Brown, 2006).

10. Harry Turtledove, *How Few Remain: A Novel of the Second War between the States* (New York: Ballantine, 1997).

11. A. Lincoln to Andrew Johnson, 26 March 1863, frame 1144, reel 2, NARA M858.

12. Robert Durden, *The Gray and the Black: The Confederate Debate on Emancipation* (Baton Rouge: Louisiana University Press, 1972).

13. *New York Times*, 26 March 1866, quoted in Glatthaar, *Forged in Battle*, 251.

14. *Glory*, dir. Edward Zwick, Columbia Tristar Productions, 1989.

Page numbers in *italics* refer to tables and illustrations.

slavery (*continued*)
24–25; as cause of war, 3–4; East coast soldiers and, 84, 86; impressment into Union army and, 4–5; as increasingly white, 32; marriage and, 28–29; northern attitudes toward, 2; Thirteenth Amendment and, 154; typical southern diet in, 55–56

slaves, as freeing selves by crossing Union lines, 2–3, 28

"slave to soldier to man" slogan, 14, 27–31, 37

slave traders, and mixed-race men, 35

Slawson, Robert, 64

smallpox: Benton Barracks and, 95–98; diagnosis of, 67; East coast and, 84; white vs. black, 11

Smart, Charles, 125, 139–40

Smith, Joseph, 15, 97

Smith, J. V. C., 59

Smith, Kirby, 121

Smothers, Samuel, 124

sources, reliability of, xvi–xvii

South Carolina, 3, 81–86

Southerners as soldiers, 30

spirometry, 51

spurious vaccination, 96

St. Louis, in Civil War, 86–88. *See also* Benton Barracks

Stanton, Edwin M.: American Freedmen's Inquiry Commission and, 28; Barnes and, 133; Fry and, 144; Saxton and, 4; Ullmann and, 105

statistics, problems with, xiii–xv

Steckel, Richard, 7

Steiner, Paul: on Benton Barracks, 91; on medical staffing, 58; mortality rates and, xi, xv; on 65th U.S. Colored Infantry Regiment, 12

Strain, Isaac, 15–19

Sumner, Charles, 33

surgeons: African American, 62–66; anatomy and, 100; food supplies and, 131; in Louisiana, 116; promotion from assistant to full, 58; promotion of hospi-

tal stewards to rank of, 59, 60–62; quality of health care provided by, 67–71, 73–79; regimental, 77; as sadistic and cruel, 71–73

Taylor, Susie King, 69–70, 78

Taylor, Zachary, 122

terminology, xiii

Texas, 120–22, 126, 132–34. *See also* scurvy epidemic in Texas

theory, comments about, xv–xvi

therapy, 55

Thirteenth Amendment, 154

Thomas, George, 75–76

Thomas, Lorenzo, 4, 60–61, 75–76

Time on the Cross (Fogel and Engerman), 7

Townsend, E. D., 60

tracking system, xiii–xiv, xv

Truth, Sojourner, 78

tuberculosis: Benton Barracks and, 102; black soldiers and, 51, 52–53; in Civil War, 56; connection between skin tone and, 147–48; East coast and, 84; in 65th U.S. Colored Troops regiment, 12; white vs. black, 11

Tubman, Harriet, 69–70, 78

Turtledove, Harry, 158

Tyler, John, 98

typhoid fever: in Louisiana, 107, 112; in northern cities, 84; in 65th U.S. Colored Troops regiment, 12

typho-malarial fever, 170n. 21

Ullmann, Daniel, 105, 106, 114

Union Army: draft of, 1; enlistment of black men in, 2–7; impressment of slaves into, 4–5; Mississippi River and, 104; numbers of black soldiers in, 6–7; physical examination at enlistment in, 145–49; racism of, 102–3; whipping and, 30, 31

Union lines, slaves freed by crossing, 2–3, 28

United States Sanitary Commission (USSC): description of, ix; B. Gould and, 151; mulatto and, 33; scurvy and, 128–29; survey of, x; Texas and, 132–34